Advances in Information Security

Series Editor
Sushil Jajodia, George Mason University, Fairfax, VA, USA

The purpose of the *Advances in Information Security* book series is to establish the state of the art and set the course for future research in information security. The scope of this series includes not only all aspects of computer, network security, and cryptography, but related areas, such as fault tolerance and software assurance. The series serves as a central source of reference for information security research and developments. The series aims to publish thorough and cohesive overviews on specific topics in Information Security, as well as works that are larger in scope than survey articles and that will contain more detailed background information. The series also provides a single point of coverage of advanced and timely topics and a forum for topics that may not have reached a level of maturity to warrant a comprehensive textbook.

Jerry Andriessen
Thomas Schaberreiter
Alexandros Papanikolaou • Juha Röning
Editors

Cybersecurity Awareness

 Springer

Editors
Jerry Andriessen
Wise & Munro
The Hague, The Netherlands

Thomas Schaberreiter
CS-AWARE Corporation
Tallinn, Estonia

Alexandros Papanikolaou
InnoSec - Innovative Secure Technologies
Thessaloniki, Greece

Juha Röning
University of Oulu
Oulu, Finland

ISSN 1568-2633   ISSN 2512-2193   (electronic)
Advances in Information Security
ISBN 978-3-031-04229-4   ISBN 978-3-031-04227-0   (eBook)
https://doi.org/10.1007/978-3-031-04227-0

This Springer imprint is published by the registered company Springer Nature Switzerland AG
The registered company address is: Gewerbestrasse 11, 6330 Cham, Switzerland

Preface

Cybersecurity is one of today's most challenging security problems for commercial companies, NGOs, governmental institutions as well as individuals. Reaching beyond the technology-focused boundaries of classical information technology (IT) security, cybersecurity includes organizational and behavioral aspects of IT systems and needs to comply to legal and regulatory frameworks for cybersecurity. While large corporations might have the resources to follow those developments and bring their IT infrastructure and services in line with the requirements, the burden for smaller organizations like local public administrations will be substantial and the required resources might not be available. New and innovative solutions that will help local public administration to ease the burden of being in line with cybersecurity requirements are needed.

In this book, we present the design of a generic socio-technical framework for cybersecurity awareness in the organizational context, building on and contributing to the European multi-level cooperative/collaborative cybersecurity approach laid out by the 2013 European Cybersecurity Strategy and reinforced by the 2020 update to this strategy. The framework is designed to both capture the socio-technical cybersecurity requirements within an organization and provides the technical capabilities for real-time monitoring of those requirements against the local organizational context, as well as against the continuously changing cybersecurity landscape. The framework also includes technical and socio-technical measures to prevent or mitigate detected cybersecurity issues in an open and collaborative way, using artificial intelligence mechanisms (AI) to derive optimal strategies for addressing cybersecurity issues in individual organizational contexts.

The book details the implementation strategy for the framework, following modular design and agile implementation techniques in order to be able to deal on the one hand with the complexities of implementing a coherent solution in the multi-stakeholder European research project context, but on the other hand to account for the incredibly fast-changing cybersecurity environment, which requires a flexible solution in order to remain effective and efficient in the long term. This strategy has allowed SMEs from all over Europe to collaborate on the implementation and of the framework, resulting in an integrated platform that is flexible in its deployment in all possible organizational contexts, while at the same time providing a coherent and intuitive user experience.

One of the strong points of the CS-AWARE framework and platform is its conceptual and operational integration within the (currently still developing) European cooperative and collaborative cybersecurity legislation, as outlined by the 2013/2020 European Cybersecurity Strategy that is gradually implemented via, for example, the network and information security (NIS) directive, or the general data protection legislation (GDPR). In its legislative efforts, the EU has recognized that the complexities of cybersecurity can only be addressed through strong cooperation/collaboration between actors and stakeholders of different sectors critical to society and economy (including their supply chains), and mechanisms need to be created to facilitate this collaboration on a national, European, and global level. But cooperation starts locally, within individual organizations that are the ones at the front lines, with their systems getting attacked and compromised by cyber-criminal activity.

The CS-AWARE framework recognizes the importance of individual organizations being actively involved in the cooperative/collaborative efforts of the European multi-level cybersecurity approach. The framework provides mechanisms for organizations to benefit from this collaboration through utilization of cyber intelligence created by relevant actors like CERTs/CSIRTs. CS-AWARE also contributes to this cyber intelligence through the abilities to share information and evidence about cyber incidents with relevant stakeholders in an effort to allow them to create more precise intelligence about new and on-going cyber threats.

As soon as the project had finished, a spin-off company was founded in 2020, so as to commercialize the developed product. The initial efforts targeted the Italian and Greek market since they were the two countries where the project pilot installations had taken place and for this reason significant effort had already been put in analyzing the specific market characteristics of these two countries. Furthermore, not only has the need for cybersecurity awareness been strengthened across the EU, but also it seems that certain aspects of it will soon be rendered mandatory according to the particulars of the revised NIS Directive, the so-called "NIS 2.0". The spin-off company is currently pursuing additional funding to further develop the CS-AWARE system and convert it into a highly competitive product. To this direction, research proposals have been submitted to relevant EU cybersecurity calls, as well as accelerator programs were pursued for obtaining the required funding.

The CS-AWARE project was funded by the European Union (Grant Agreement number 740723), as part of the Horizon 2020 program. For researchers, this is a unique and generous opportunity to work together in an international context on ambitious goals. Funding is hard to obtain, however, as competition is high, and because the constraints and requirements of such programs tend to increase with every new program. This can be illustrated by a quote describing how the European Union sees the goals of Horizon 2020:

> Horizon 2020 is the financial instrument implementing the Innovation Union, a Europe 2020 flagship initiative aimed at securing Europe's global competitiveness. Seen as a means to drive economic growth and create jobs, Horizon 2020 has the political backing of Europe's leaders and the Members of the European Parliament. They agreed that research

is an investment in our future and so put it at the heart of the EU's blueprint for smart, sustainable, and inclusive growth and jobs. By coupling research and innovation, Horizon 2020 is helping to achieve this with its emphasis on excellent science, industrial leadership, and tackling societal challenges. The goal is to ensure Europe produces world-class science, removes barriers to innovation, and makes it easier for the public and private sectors to work together in delivering innovation. Horizon 2020 is open to everyone, with a simple structure that reduces red tape and time, so participants can focus on what is really important. This approach makes sure new projects get off the ground quickly—and achieve results faster. The EU Framework Programme for Research and Innovation will be complemented by further measures to complete and further develop the European Research Area. These measures will aim at breaking down barriers to create a genuine single market for knowledge, research, and innovation. (Source: https://ec.europa.eu/programmes/horizon2020/what-horizon-2020)

It says it all. How such political impact-driven ambitions influence the collaboration and the outcomes of an innovation project is an interesting question, and the answer can only be complicated and multidimensional. From the side of the funding agency, qualified EU-officials have the task of supervising and moderating these projects and are responsible for selecting and supporting competitive proposals that provide, all together, significant contributions to the success of the Horizon 2020 program. For selected proposals, the efforts needed to undertake innovative design projects are huge, as such projects involve multiple agents (designers, computer scientists, learning scientists, various administration professionals, students, decision-makers, etc.) for a period that generally ranges between 2 and 4 years. CS-AWARE was ambitious, because, as can be seen in this book, in many phases it required explicit collaboration between various stakeholders. CS-AWARE cannot be tailored to user needs without participation of LPA professionals in several workshops, where they explain in great detail their system network, and the way it is used. CS-AWARE cannot be implemented without local support (although we are aiming for lightweight versions), requiring collaboration between local technicians, their managers, and computer scientists. Conceptually, CS-AWARE is designed by extensive collaboration between social and computer scientists. Deployment scenarios include understanding user activity and feedback, which requires collaboration with users on a regular basis. All of this is described in the various chapters of the book. The chapters have been written by authors with different backgrounds, which gives this book its multidisciplinary flavor, characteristic of innovation projects in the EU context.

Therefore, we expect this book to be interesting to a great variety of readers: computer scientists, social scientists, public management, and to the general audience of people interested in current thinking about cybersecurity awareness.

The Hague, The Netherlands   Jerry Andriessen
Tallinn, Estonia   Thomas Schaberreiter
Thessaloniki, Greece   Alexandros Papanikolaou
Oulu, Finland   Juha Röning

Acknowledgment

On behalf of all authors in this book, the editors like to express their regards to the European Union, for the funding of the CS-AWARE project.

Contents

List of Contributors

Jerry Andriessen Wise & Munro, The Hague, The Netherlands

Arianna Bertollini Municipality of Rome, Rome, Italy

Claudio Ferilli Municipality of Rome, Rome, Italy

Massimo Ferrarelli Municipality of Rome, Rome, Italy

John Forrester Cesviter Consulting, Rome, Italy

Kim Gammelgaard RheaSoft, Aarhus, Denmark

Manuel Leiva Lopez Cesviter Consulting, Rome, Italy

Christian Luidold University of Vienna, Vienna, Austria

Alexandros Papanikolaou INNOSEC, Thessaloniki, Greece

Mirjam Pardijs Wise & Munro, The Hague, The Netherlands

Omar Parente Municipality of Rome, Rome, Italy

Thanasis Poultsidis Municipality of Larissa, Larissa, Greece

Gerald Quirchmayr University of Vienna, Wien, Austria

Juha Röning University of Oulu, Oulu, Finland

Thomas Schaberreiter CS-AWARE Corporation, Tallinn, Estonia

Massimo Della Valentina Cesviter Consulting, Rome, Italy

Christopher Wills Caris Partnership, Fowey, UK

Chapter 1
A Case for Cybersecurity Awareness Systems

Thomas Schaberreiter, Gerald Quirchmayr, and Alexandros Papanikolaou

1.1 Introduction

In this Chapter we present an introduction to the increasingly challenging topic of cybersecurity in general, and the specific case of cybersecurity in local public administrations (LPAs). We outline the current cybersecurity landscape, as framed by recent European advances in cybersecurity law and regulation like the EU cybersecurity strategy, the network and information security directive (NIS) and the general data protection regulation (GDPR). Furthermore, an analysis of current trends in the global cybersecurity threat landscape is provided, and we give an overview of the currently available cybersecurity solutions for organizations on the technological, organizational and trans-organizational (national, European and global) collaborative level. We discuss the advantages/disadvantages of each solution and we will argue that all of those elements fulfil an important role, but in order to adequately protect the data managed by LPAs, a more holistic, socio-technical approach that is built upon cybersecurity awareness and collaboration in the organizational context is required.

We discuss the relevance of the European and global cybersecurity environment for the specific context of local public administrations and the and applicability of different approaches. Our analysis has shown that the data which is managed by

T. Schaberreiter
CS-AWARE Corporation, Tallinn, Estonia
e-mail: thomas.schaberreiter@cs-aware.com

G. Quirchmayr
University of Vienna, Wien, Austria
e-mail: gerald.quirchmayr@univie.ac.at

A. Papanikolaou (✉)
INNOSEC, Thessaloniki, Greece
e-mail: a.papanikolaou@innosec.gr

LPAs (both citizen services and organizational services) are the most critical asset to be protected. We will outline the importance of protecting this data through the entire information flow caused by the day-to-day operations and processes of an LPA workflow. Following up on those aspects, we show how cybersecurity awareness in the organizational context can bring additional and novel elements to the existing state-of-the-art solutions, and which concepts, methods and methodologies are required to implement such a solution. The core concept includes continuous monitoring of system elements through dynamic, data driven risk and incident management. This requires the introduction of intelligent data analysis systems to provide accurate and context specific monitoring results and awareness. Furthermore, in order to fully utilize the advantages of cybersecurity awareness, the organizational culture needs to shift as well, from seeing cybersecurity as an individual or IT department task, to seeing cybersecurity as a collaborative socio-technical effort that requires the contribution of each individual in the organization - supported by a technological platform that allows this collaboration. Finally, we illustrate how good cybersecurity awareness in an organization lays the foundations for more advanced features like system self-healing (awareness of a problem is the basis for automated implementation of prevention or mitigation mechanisms) and cybersecurity information sharing (exact determination of the cause of an incident through awareness allows to determine information to be shared with cybersecurity communities for collaborative cybersecurity efforts, or in the context of mandated incident sharing requirements, as for example imposed by NIS or GDPR).

In this chapter, first an overview of the current cybersecurity landscape is given in Sect. 1.2. State-of-the-art cybersecurity practices are discussed in Sects. 1.3 and 1.4 details the currently actively evolving European legislative framework concerning cybersecurity. Section 1.5 highlights the cybersecurity requirements of LPAs, and Sect. 1.6 presents the case for advanced cybersecurity awareness solutions in this context. Finally, Section 1.7 provides a summary and gives an outlook on the following Chapters of this book.

1.2   The Cybersecurity Landscape

Over the past decades we have seen a steady rise in both system vulnerabilities, and in attacks exploiting them. These unfortunate developments are well-documented by ENISA (2019) and EUROPOL (2021) in their cyber threat reports. Accordingly, the European Union has in its 2013 Cyber Security Strategy set the frame for a coordinated strategic approach to addressing the issue (European Commission, 2013). The currently most important legislative cornerstones based on this strategy are the GDPR (European Parliament, 2016b) and the NIS directive (European Parliament, 2016a) both of them having a significant impact on the perception of the importance of cyber security for European society, infrastructure and economy. Recent information operations aimed at destabilizing institutions and organizations are documented by The European Centre of Excellence for Countering Hybrid Threats (Hybrid CoE, 2021).

While targeted attacks on Local Public Administrations (LPAs) are still the exception, the danger of an LPAs infrastructure being used as backdoor for intruding more important governmental systems is very real (Coppolino et al., 2018). As research on supply chain security shows, this type of threat keeps increasing (Lamba et al., 2017) and becomes especially relevant for LPAs that manage critical (information) infrastructures. While ransomware attacks in 2017 (Hsiao & Kao, 2018) have increased the awareness for this problem, the threats posed by malware, DDos and APT attacks still tend to be underestimated. With IoT devices now becoming omnipresent, LPAs can easily become collateral damage in attacks that get out of control or that are launched in a massive way against potentially ill-defended targets of opportunity to exploit unpatched systems suffering from unattended vulnerabilities. These attacks have in the past caused and continue to cause large-scale impacts in an organization if this vulnerability is present. The embarrassingly clear reason for an attack being successful is all too often outdated software (due to limited resources) or unpatched software due to unawareness of update practices or slower response times in update practices due to resource limitations. Seen against the background of the current 2019 IOCTA (Internet Organised Crime Threat Assessment) report, the potentially devastating consequences of such negligence become evident (Europol., 2019):

- Ransomware remains the top cyber threat in 2019, and phishing and vulnerable software (especially vulnerable remote desktop protocols) remain the main infection vectors.
- Data is the key target for cybercrime.
- There is a growing concern in organizations about sabotage, for example by destructive ransomware.
- A growing need for collaboration between network and information security efforts and law enforcement is identified.

With the value of data managed by LPAs becoming an increasingly important asset, the probability of attacks raises continuously and it may therefore not be a question of "if", but rather of "when" an attack occurs. Consequently, in LPAs the impacts of such attacks need to be limited to protect the managed data. This is best achieved through an awareness system that monitors relevant high-value assets for security relevant parameters like outdated software components or unusual behaviour that would indicate an attack. As past cases have shown (GovCERT.ch, 2016), one of the major challenges is the early warning against and the early detection of vulnerabilities, related exploits and ultimately attacks launched against an LPA. In such an environment, keeping an up-to-date situational awareness picture of an LPAs cyber infrastructure becomes a high priority. To achieve this, a more user centric approach is required that allows LPAs to be aware of their systems security state and the state of the data managed by those systems at all times, but at the same time facilitates collaboration between actors within the organization (like management, IT, service users), but also actors from outside the organization like NIS authorities or law enforcement. The remainder of this Chapter will outline the current state-of-the-art cybersecurity mechanisms at the disposal of LPAs to ensure cybersecurity, and we will identify the gap in the current state-of-the-art that needs to be addressed in order to achieve a collaborative approach to cybersecurity based on cybersecurity awareness in LPAs.

1.3 The State-of-the-Art in Cybersecurity Practice

Digital and web-based technologies have transformed the way citizens can interact with LPAs, by facilitating the provision of electronic and remote services. In this way, citizens can be served more quickly and in a more efficient manner, since some of their requests can be processed without having to physically go to the LPA. Nevertheless, the provision of such services, as well as the involved data (both in transit and at rest) require a sufficient level of protection against cyberattacks, since a single breach can affect the data of many citizens, which usually is of significant sensitivity. In the remainder of this Section, various solutions for protecting LPAs will be presented, from different perspectives (technological, organisational, national/European/global).

1.3.1 Technological

When it comes to adequately protecting an LPA against cyberthreats, it is necessary that certain security mechanisms are installed so that a satisfactory cybersecurity protection level can be achieved. Firstly, a perimeter defence must be set up, which serves as the first line of defence. The primary security mechanism used for this purpose is a firewall that essentially divides the network into internal (trusted) and external (untrusted) sections and monitors the network traffic flows between them. Through the firewall's rules the organisation's security policy is implemented and according to them the communication requests are allowed or dropped.

Should remote connectivity for external users be required, the use of a Virtual Private Network (VPN) is usually applied. External users connect to the firewall and authenticate themselves. Once this process is completed, a secure connection between the external user's device (e.g. computer) and the organisation's firewall is established, the so-called VPN, which grants the authenticated user access to the organisation's internal network.

The internal network also requires a sufficient degree of protection, to both prevent e.g. malware infections from happening and prevent them from spreading across the whole internal network in the unfortunate case that they manage to bypass the installed security mechanisms. For instance, a user looking for something on the Internet may unintentionally initiate a malware download to the local computer, thus creating the initial infection, that may spread to the rest of the network if there is no mechanism to contain it. This kind of protection is usually offered via endpoint protection software, which consists of a collection of antivirus, firewall and access control features (e.g. permitting the connection of USB storage devices or not).

Once the aforementioned security mechanisms have been installed, it is equally important to monitor the activity in the organisation's information system, aiming to detect any suspicious behaviour as soon as possible and consequently react promptly to deal with the detected issue. A popular mechanism for performing such a task is an Intrusion Detection/Prevention System (IDS/IPS), which monitors the network traffic at designated points of the organisation's network, analyses the traffic and detects potentially suspicious behaviour, according to the rules that have been set to it. An IPS usually interacts with a firewall, so that it can react to a detected threat (e.g. block the source IP address of the detected suspicious traffic).

A security mechanism similar to the IDS/IPS is a SIEM system (Security Information and Event Management). Their main difference is that the IDS/IPS monitors network traffic, whereas the SIEM analyses in real time the activities in an IT environment (namely, log files that contain information about various events happening at key network nodes). Any detected abnormal activity raises an alert, which has to be further investigated by the system administrator to determine whether this may be an indicator of a cyberattack taking place (e.g., too many user log-in failures may be due to a password-guessing attack).

Nevertheless, both IDS/IPS and SIEM solutions require administrators to monitor the alerts they produce and look into the involved issues. In several cases investigating the issue may require significant time and cybersecurity expertise that the majority of system administrators may not have. For this reason, there is a trend for organisations to outsource their network security services to a Managed Security Services (MSS) provider, who offers 24/7 monitoring and response by their team of cybersecurity experts. Of course, a 24/7 service has quite a significant cost associated with it, which does not make it a very attractive option for small organisations with a tight budget.

1.3.2   Organisational

Installing security mechanisms is not enough for achieving a high cybersecurity level, especially as the size and the complexity of an organisation increases. Organisational risk management is the process of identifying, assessing and controlling threats, which contributes to the enhancement of its cybersecurity level. The most important assets are initially identified, together with applicable threats to them and suitable controls are put in place to keep the calculated impact of a potential mishap as low as possible.

The need for organisational measures is also covered by the well-established family of the ISO27000 standard that involves the security of an organisation's information systems. Compliance to this standard (certified by appropriate certifications obtained via audits) implies that suitable procedures have been put in place to conduct information security management according to the best practices.

Common Criteria (CC1, 2017, p. 1; CC2, 2017, p. 2; CC3, 2017, p. 3) is a certification through which computer systems users can specify their security functional requirements (SFR) and security functional assurance requirements (SARs) using protection profiles (PP). Technology vendors can then hire testing laboratories to evaluate their products to determine whether they meet the said requirements. The European Agency for cybersecurity, ENISA, has recently published the candidate version for a European certification scheme (ENISA, 2020) to cope with up-to-date cybersecurity requirements.

Nevertheless, using security standards in an organisation has certain drawbacks that in some cases may render the implementation of such a standard not worthwhile. To start with, they require significant expertise and familiarisation for both implementing and maintaining them, which effectively translates into a need for significant additional resources. This includes the training required for familiarising the personnel with the newly-introduced procedures, as well as the ongoing training for creating awareness. Furthermore, the various procedures that will be introduced as part of the standard's implementation will impose an additional overhead to the existing daily activities. It may therefore not be practical for relatively small organisations, as they are lacking resources to handle this extra overhead. The financial aspects should also be considered too, as the implementation of such standards may require special equipment that would have not been bought otherwise, or that has to meet certain specifications and is therefore more costly than lower-level alternatives. What is more, the cost of obtaining and maintaining a relevant certification should also be considered.

1.3.3   National, European and Global Cybersecurity Efforts

At the European level, the EU Cybersecurity Strategy (European Commission, 2013) suggests a series of actions to enhance the cyber resilience of IT systems, reduce cybercrime and strengthen EU international cyber security policy and cyber defence. The cybersecurity strategy states that Computer Emergency Response Teams (CERTs) and Computer Security Incident Response Teams (CSIRTs) are responsible for coordinating and supporting the response to a computer security event or incident. They exist at national, European and international level, and they communicate hierarchically so as to maximise their effectiveness in cybersecurity protection (e.g. the European CSIRT network[1]).

The legal and regulatory frameworks are also to be considered. For instance, in Europe, the GDPR (European Parliament, 2016b) has quite recently come into effect, a European Regulation for unifying the protection and privacy of EU

---

[1] https://www.enisa.europa.eu/topics/csirts-in-europe/csirts-network.

citizens' data. In order to achieve compliance with the GDPR both technical and organisational measures need to be utilised. Once a satisfactory level of compliance has been achieved, the organisation's cybersecurity level is also improved.

The NIS Directive (European Parliament, 2016a) also aims to boost the overall level of cybersecurity in the EU. At national level it requires Member States to have national CSIRTs, perform cyber exercises etc. to achieve and maintain a high cybersecurity level. It also emphasises on the need for cross-border collaboration between EU countries as well as on a global level, through their CSIRT network and other competent groups and task forces. Finally, it requires Member States to supervise their critical infrastructure sectors (energy, transport, water, health, digital infrastructure and finance) and adequately protect it against cyberthreats, for example by establishing national competent authorities (CAs) to collaborate on risk assessment with the operators of critical infrastructure. Such a supervision can be performed by suitable cybersecurity mechanisms (e.g. SIEM) that will be installed in the organisation's information system. There are cases where such critical infrastructure services are controlled by LPAs or lie within the competency of a larger LPA. Hence, monitoring of such infrastructures will certainly help in creating cybersecurity awareness, as system administrators will be informed about detected cyberthreats that previously used to go unnoticed. What is more, the information about such cyberthreats could further be exploited by being forwarded to competent CERTs/ CSIRTs and therefore promote cybersecurity awareness at a national and/or European level.

A closer look at those legislative efforts in Europe and its implication on organisational cybersecurity risk and incident management, as well as a brief outlook on relevant proposed legislation currently on the horizon will be given in Sect. 1.4.

1.4   European Union Legislation Driving a Paradigm Shift

Starting with the EU Cybersecurity Strategy in 2013, the legal environment concerning cybersecurity has seen significant changes in Europe, with the implementation of the GDPR and the application of the NIS Directive resulting in respective national legislation in EU member states being considered as the major cornerstones established between 2016 and 2018. Besides those two important legislations that aim to improve cybersecurity on the operational level, the EU Cybersecurity Act was introduced as a regulation in 2019 (European Parliament, 2019), strengthening the European position on cybersecurity by expanding the roles, responsibilities and resources of the European cybersecurity agency ENISA, and by introducing the legal basis for an EU wide cybersecurity certification framework for products, services and processes in the context of information and communication technology, making sure security can be certified in a uniform way across Europe.

The EU wide certification framework follows a risk-based approach that uses assurance levels (basic, substantial, high) to quantify the level of risk associated to a product, service or process. The assessment will be based on a comprehensive set of rules, technical requirements, standards and procedures. While the EU cybersecurity act offers the guiding principles for the certification framework, the implementation and application of the framework is guided by the European cybersecurity certification group (ECCG), with ENISA being tasked to prepare candidate schemes for the framework itself.

The GDPR, which has since its introduction in 2016 achieved major goals, such as raising awareness of the importance of privacy protection in industry, organizations, and with citizens, contains in some of its articles partially new, far more precise sets of obligations and recommendations than previous European legislation. Most notably these address the safeguarding of the following rights and principles:

- Right of data portability
- Right to be forgotten
- Privacy by default
- Privacy by design
- Risk-based approach to privacy protection
- Effectiveness of security mechanisms in place to be monitored and controlled

While this does heavily impact system design and development, especially the obligations to take a risk-based approach to privacy protection and to ensure that the effectiveness of security mechanisms in place are monitored and controlled result in the need to look at the operational security of systems in a different way. Traditionally, a vulnerability-centred information security management approach was pursued, and the focus was mostly on safeguarding systems and their components. Risk-based privacy protection demands a radical shift towards an information asset-oriented view and the obligation to monitor and control the effectiveness of security mechanisms results in a change from snapshot type audits to continuous auditing. As this quite radical change has effects way beyond the technological infrastructure, a more holistic and integrated view needs to be supported, comprising people, processes, and technology. For such an approach being successful, a new way of looking at the protection of information assets is needed, including risk-oriented thinking, situational awareness, and a continuous evaluation of the protection measures deployed.

This paradigm shift towards a more holistic, risk-oriented, and situational awareness focussed approach is further strengthened by the NIS Directive. The mandated reporting of relevant cyber incidents requires, like the data breach notification obligations in the GDPR, situational awareness and the underlying monitoring and analysis capabilities. The requirement to implement adequate technological and organizational safeguards for the protection of critical infrastructures also leads to an increased need for providing cyber situational awareness for information systems operators and for planners.

The NIS Directive has alone through its coming into effect made cybersecurity a core item in the management of critical infrastructures. The sectors identified as critical infrastructure by the NIS directive to provide essential services are the following: Energy, Transport, Banking, Financial market infrastructures, Health, Drinking water supply and distribution, Digital infrastructure. This means that essential services offered in these areas are covered by the NIS directive. A further specialised piece of legislation was introduced with the 2018 European Electronic Communications Code (European Parliament, 2018) to cover cybersecurity related to the telecommunications sector following the approach of the NIS directive.

While the most immediate, and therefore intensively discussed effect is felt in the obligation to report significant cyber incidents, according to Article 1 (1) the overall the goal "of achieving a high common level of security of network and information systems within the European Union" leads to a whole series of efforts, as listed in Article 1 (2):

(a) lays down obligations for all Member States to adopt a national strategy on the security of network and information systems;
(b) creates a Cooperation Group in order to support and facilitate strategic cooperation and the exchange of information among Member States and to develop trust and confidence amongst them;
(c) creates a computer security incident response teams network ('CSIRTs network') in order to contribute to the development of trust and confidence between Member States and to promote swift and effective operational cooperation;
(d) establishes security and notification requirements for operators of essential services and for digital service providers;
(e) lays down obligations for Member States to designate national competent authorities, single points of contact and CSIRTs with tasks related to the security of network and information systems.

While most of the agenda items listed above primarily target the national legislation and governments, especially (d) has very direct impacts on operators of essential services and on digital service providers. These are then described in further articles, with Article 16 "Security requirements and incident notification" being the central reference for implementing security measures. Mandating a risk-oriented approach in Article 16 (1),[2] it introduces the requirement to consider "(a) the security of systems and facilities; (b) incident handling; (c) business continuity management; (d) monitoring, auditing and testing; (e) compliance with international standards." Well-structured cybersecurity frameworks, such as the ISMS suggested in the ISO/IEC 27001:2013 standard (ISO, 2013) and the NIST/CSF (National Institute of Standards and Technology, 2018) do in this context prove their high value.

It is especially the obligation to implement a risk-aware monitoring, auditing, and testing which results in the need to expand existing solutions towards risk-awareness and therefore also threat intelligence and continuous monitoring and auditing. For service providers running or servicing critical IT systems for LPAs and for LPAs themselves, this means a radical shift from a traditional technically oriented bug-fixing approach to a risk-aware process.

---

[2] Having regard to the state of the art, those measures shall ensure a level of security of network and information systems appropriate to the risk posed [...] (Article 16 (1) NIS Directive).

The data breach and incident reporting obligations in the GDPR[3] and in the NIS Directive[4] do both require system operators to be able to detect incidents, analyse, and categorize their impact. These obligations do in turn require a well-developed situational awareness capability. With IDS and SIEM systems already in place in many organizations, the next logical step towards achieving this capability is the provisioning of a situational awareness layer built on top of continuous monitoring and assessment processes.

As Gregory J. Touhill states in chapter three of his book 'Cybersecurity for Executives: a practical guide' (Touhill & Touhill, 2014): "*Cybersecurity is primarily about risk management*". To meet the requirements of taking a risk-oriented approach, and to continuously monitor and assess the effectiveness of protective measures, cyber threat intelligence becomes a must. Situational cybersecurity awareness consequently needs to provide capabilities to continuously monitor, assess, and audit internal systems, to monitor threat developments, and to map these two views to identify and assess risk.

This paradigm shift driven by European legislation is still on-going and evolving. In 2020 the EU published a new and updated cybersecurity strategy (European Commission, 2020a) in an effort to update the strategic direction for addressing new cybersecurity challenges emerging since the publication of the original cybersecurity strategy in 2013. It was identified that a stronger cross-sector and cross-border collaboration within the EU is required in order to be able to ensure cybersecurity throughout the supply chain. The major challenges addressed by the strategy include:

- Increasing (cross-sector and cross-border) interdependency of European critical infrastructure
- Increasing geopolitical tensions, including control over technology across the supply chain
- Increasing criminal and malicious activity against critical infrastructure and on-line services
- The lack of collective situational awareness of cyber threats in the EU

The strategy calls for even stronger cooperation/collaboration between actors and stakeholders. The necessity of shared situational awareness in this context becomes even more apparent as a key enabler to address those challenges. In concert with the updated cybersecurity strategy, two proposal directives were published that (a) propose an update to the 2016 NIS directive, the NIS2 (European Commission, 2020a) and (b) propose a directive on the resilience of critical entities (European Commission, 2020b)which essentially aligns the framework for dealing with physical risk to critical infrastructure with the framework outlined by the NIS/NIS2 directives.

---

[3] GDR Article 30 "Notification of a personal data breach to the supervisory authority"; GDPR Article 31 "Communication of a personal data breach to the data subject".

[4] NIS Directive Article 14 "Security requirements and incident notification".

The proposed NIS2 directive builds on the success of the NIS directive, and implements measures to address the challenges that have been identified in the first years of application of the NIS directive. While the NIS mechanisms have proven effective in a sectoral and national context, it has been identified that the lack of cross-sector and cross-border collaboration prevents the NIS from unfolding its full potential. Some of the most important aspects proposed to address this aspect in NIS2 include:

- In addition to the seven critical sectors listed by the NIS (Energy, Transport, Banking, Financial market infrastructures, Health, Drinking water supply and distribution, Digital infrastructure), 3 additional sectors have been added: Waste water, Public administration and Space.
- To account for the increasing relevance of the supply chain for cybersecurity, the NIS2 introduces an additional class of important sectors, facing the same cyber-security management and reporting requirements as critical sectors (but a different way of penalising non-compliance). The list of important sectors includes: Postal and courier services, Waste management, Manufacture, production and distribution of chemicals, Food production, processing and distribution, Manufacturing, Digital providers
- The classification of critical entities will not be a member state obligation in NIS2, as it has caused cross-border consistency issues. In NIS2, any organisation operating in a critical/important sector listed by the directive will be obliged to be compliant with the directive, with the exception of micro and small entities.

The proposed directive on the resilience of critical entities is designed to replace and significantly extend the scope of existing legislation on critical infrastructure protection (Commission of the European Communities, 2006). The scope of the directive is to provide a framework to deal with physical risk against critical infrastructure in Europe. In line with the 2020 European cybersecurity strategy, it proposes the same framework as outlined by the NIS2 directive for managing cyber risk, applying it to the management of cyber risk. This includes an alignment in the identification of critical and important entities. Aside from being able to benefit from the same level of cross-sector and cross-border collaboration fostered by the NIS2 directive, the alignment in managing cyber risk and physical risk accounts for the increasing concerns over cyber-physical threats, where incidents in the cyber domain can have physical consequences and vice versa. The alignment of the two frameworks will allow for stronger cooperation and collaboration in this context.

All in all, the European cybersecurity legislation, both regarding security and privacy, are based around a systemic-holistic understanding of the problems causing cybersecurity issues, and an increasingly strong cooperation/collaboration on solving those problems. Situational awareness is at the core of all of those efforts.

1.5   Cybersecurity Requirements for LPAs

Like any organisation of a certain complexity, LPAs vary significantly in their individual set-up and the socio-technical interrelations that evolve naturally guided by organizational culture and policies. Looking at it from the European perspective, the set-up and structure of LPAs is also heavily influenced by national legislations, so there are significant national differences in how LPAs operate based on the governing framework that defines the responsibilities and obligations of LPAs. All of those aspects influence cybersecurity, and need to be taken into account when devising cybersecurity solutions. While based on those environmental conditions it becomes clear that a definitive and generalizable characterization of cybersecurity requirements for LPAs seems unrealistic, this Section will give a guiding overview of how IT systems are usually involved in providing administrative services to citizens and businesses in its constituency - and what this means for cybersecurity. In the following Section we will show how the individual socio-technical organizational differences are a crucial aspect in cybersecurity considerations, and how CS-AWARE was designed to be able to embrace this fact and provide cybersecurity solutions tailored to individual organizational cybersecurity needs.

There are two general distinctions in IT services operated within LPAs: (1) Services that are offered to citizens and (2) services to manage operations within the LPA. The first category includes simple services that are offered to inform citizens, like the web page or keeping and providing public records, as well as complex e-government services that allow to interact citizens or organizations with the LPA, for example to apply for permits. Most of those complex services map to strictly defined business processes that are not completely automated and require back-office handling and approvals, often by several individual departments in the LPA. The second category includes the more standardized organizational administrative operations, like for example human resources (HR) and payroll systems. While both service categories, when looking at them strictly from the technological standpoint, follow similar implementation and deployment patterns, there are two relevant distinctions that are important from the cybersecurity perspective: First, the data managed by the services in each category is fundamentally different in the sense that one category only manages and processes data (both sensitive and personal data) of employees, while the in the other category potentially sensitive and/ or personal data about every citizen is managed and processed, and a breach in those systems would potentially leak a significantly larger amount of data. The second major distinction is in the potential complexity of business processes that are the basis for the offered service. While organizational management usually follows fairly standard organizational practices not unique to LPAs, IT-based or IT-supported administrative services for citizens may require the implementation of complex processes, often following the same policies and procedures implemented for the offline counterpart - including integrated processes where IT only supports parts of the process like communication and data gathering, while the processing is still carried out as a back-office task. In terms of cybersecurity, this increases the potential attack points and thus the requirements in cybersecurity.

One additional trend that could be observed in recent years is the continuing centralization of services that traditionally have been the responsibility of LPAs, but due to the advantages of digitalization can now be managed directly at the national level. One good example for this is the citizen registry, which has - especially in European countries - traditionally been (at least partially) a local or regional responsibility, but is now increasingly managed as central register. From the cybersecurity perspective, this development reduces the burden on LPAs to provide adequate protection for the services, but at the same time limits the control about how the service and data are managed based on the individual needs to the LPA, which may have implications on cybersecurity, e.g. in the way how access control to the service and data is handled.

This quick look at how IT services are usually utilized in LPAs makes one thing obvious: No matter if the service is a citizen or an organizational management service, the most critical asset of LPAs are the data managed and processed by those services. The protection of the data in all stages of the day-to-day processing of the data (including data-in-storage and data-in-transition) needs to be at the core of all cybersecurity considerations. This is, especially in Europe, also very much in line with current legal and regulatory efforts like the GDPR or NIS directive, which apply to local public administrations in the context of data protection and - in many cases - as providers of essential services. To ensure legal compliance, a deep understanding of systems, interactions and processes for the identification, monitoring and mandated reporting of incident and/or breach information is required.

In this sense, a strong cybersecurity requirement in LPAs needs to be a holistic understanding of the IT systems (e.g. server, database, service, network, ...) and the interactions between them in a socio-technical way to understand how the organizational structure can influence the security of e.g. the data that is transmitted, processed and stored by those systems. This includes the identification of the key business processes in day-to-day operations that define the main services provided by an LPA, and how those processes cause data flows and data transformations throughout the systems. The human factor plays a major role in those interactions by managing or administering processes, services or systems, which can have major implications for security. This effort may be a manageable task in small to medium-sized LPAs (number of services, number of users and support personal is limited), but can be a major undertaking in metropolitan areas, where the complexity of the systems may seem overwhelming. Furthermore, the larger the municipality, the more external suppliers that are part of the system play a relevant role in the process.

While the requirement for a holistic understanding of one's systems seems intuitively clear, the reality - especially in larger organizations - follows a more compartmentalized approach. The proper and adequate security mechanisms are put in place on each system level individually (e.g. network, service, application, ...), without a deep understanding of how this affects or relates to dependencies. While most of the times this has no negative impact on the security, it tends to lead to an unnecessarily complex security architecture, and each small change in the system set-up may require a coordinated change in multiple security appliances, like white-listing an IP in a multitude of firewalls if a new client is configured. In general, LPAs usually follow cybersecurity best practices, like applying updates regularly, installing firewalls and anti-virus software and, if applicable intrusion detection/prevention systems (IDS/IPS). Due to limited resources in the public sector, investment in cybersecurity that goes beyond the best practice is often not feasible. For example, procurement of new equipment to replace legacy systems is often not possible due to budget constraints. This can cause security issues if the components are beyond the vendor specified support cycle. Similarly, systematic cybersecurity awareness programs to inform employees about cybersecurity issues and mitigation possibilities are the exception rather than the norm. Awareness usually happens reactive rather than proactive - if an issue happened and is fixed by the IT department, the user will be made aware of the cause (e.g. not to click on suspicious email attachments).

1.6   Cybersecurity Awareness in the Context of an LPA

We have seen from the previous Sections that LPAs have a specific requirement to protect data, and which common security mechanisms on technical, organizational and legal/regulatory level are available to achieve this task. However, we have also seen that few of the existing cybersecurity measures for organizations take a holistic view on protecting their organizational assets, and that especially the protection of data that traverses many boundaries in day-to-day operation within an LPA may not be covered in current state-of-the-art cybersecurity considerations. In order to address this shortcoming, a new approach of how organizations manage cybersecurity knowledge is required. The key is to achieve a holistic understanding of an organization's cybersecurity status at each level in the organization: collaborative organizational cybersecurity awareness.

To achieve this goal, two major aspects need to be addressed:

- Cybersecurity needs to be seen as a collaborative effort within the organization, and each department/employee - not only the IT department - needs to feel the responsibility and have the skills to address cybersecurity according to their role in the organization. This necessitates that a certain level of awareness and training about cybersecurity threats and how they relate to the organizational context is available in each relevant department, and that a certain level of collaboration between stakeholders within the organization, as well as with external expert communities (e.g. CERTs/CSIRTs) is possible.
- The ability to monitor and assess the security state of key assets within the organization in order to be able to detect cybersecurity issues and attacks in real-time, and have the ability to assess and react in a proactive way. Dynamic and data-driven risk and incident management is a way to achieve situational awareness within an organization, and the continuing advancement of artificial intelligence

(AI) to detect abnormal behaviour in large data sets is a key enabling technology in this context.

Both of these goals within an organization are supported by the current shift in the cybersecurity landscape – especially in Europe: Driven by the European cybersecurity strategy of 2013/2020 (with the NIS and GDPR being examples for legislation stemming from the strategy), a cybersecurity environment that centres around cooperation and collaboration among key actors in cybersecurity (Network and Information Security actors, Law Enforcement Agencies, Defence, Industry and Academia) is implemented with the aim to gather and analyse cybersecurity intelligence in the respective areas of responsibility, and share this intelligence among the communities and the public. In the context of network and information security, the NIS obliges organizations that are providers of essential services to share information about cyber incidents with the relevant authorities, and the GDPR obliges all businesses handling personal data to share information about any data breach.

A core question is as to how this does relate to the organizational context and the goals described above. The threat intelligence created and shared by relevant cybersecurity communities based on collaborative knowledge can be utilized by organizations in order to better understand the current cybersecurity landscape. This can range from a better awareness of long-standing and recurring issues and threats, like the statistical analysis of top threats and their mitigation mechanisms, to highly dynamic threat intelligence, like the constant identification of servers and IP addresses used by malicious actors to be able to apply mitigations to such issues in near real-time. This information can be used to facilitate the collaborative cybersecurity efforts and awareness within the organization, by utilizing context specific information shared by cybersecurity communities within the relevant departments and by the employees for better awareness and preparedness. On the other hand, the legal requirement to share cybersecurity information in case of incidents requires organizations to better understand the cybersecurity implications within their own systems, what damage an attack can cause and what data sources within the systems (e.g. log data on various system levels) can be utilized to detect incidents and, in the worst case, allows forensic analysis of the damage caused by an attack to be shared with the authorities. This in turn raises the level of cybersecurity knowledge within the organization, and creates a basis for more advanced technological systems to build on this understanding and apply automated solutions for better awareness and to ensure legal compliance.

In order to make use of this potential, future cybersecurity efforts for organizations like LPAs need to focus on building supporting technology for the data driven processes described above. This includes a better utilization of the data sources (both those that are able to describe the security state of organizational systems, and those that describe the global cybersecurity state), using AI technologies to automatically detect suspicious behaviour indicated by organizational data sources, identify and correlate it with information provided by competent authorities, and create awareness for the appropriate departments and employees within the organization by distributing the information about incidents and potential prevention/

mitigation strategies in a way that is appropriate to the profiles of the individual employees.

At the same time, the individual socio-technical set-up of each organization, and its implications for cybersecurity, need to be better understood and utilized by AI based cybersecurity systems. The digital footprint of an organization is not only left by its IT components, it is created by the people that interact with the systems and the business processes they conduct utilizing the IT. While a good AI system works extremely well to identify abnormal behaviour in data sources, and has the ability to learn and adapt over time - especially if large data sets are available, the daily routine of an organization often does not follow strict and deterministic behaviour patterns - causing false positive detection of abnormal behaviour. Especially in the context of learning cybersecurity related behaviour patterns, there is much potential in combining specific organizational knowledge to help AI algorithms to better understand individual behaviour and improve detection accuracy. This goes in line with the organizational need for collaborative cybersecurity, and the need to better understand its assets, dependencies and their relevance for cybersecurity described above. This knowledge can be harnessed as an input for training AI algorithms.

Building on the increased organizational cybersecurity awareness that collaboration in the organization and the ability to monitor cybersecurity in an automated way are providing, a baseline security concept is established that can be an enabling factor for more advanced cybersecurity features. In this work we will have a look at two such features building on cybersecurity awareness, namely system self-healing and cybersecurity information sharing.

System self-healing is a concept that allows to prevent or mitigate cyber incidents in an automated way, by for example changing system configuration or applying patches to vulnerable software in case continuous monitoring has detected an issue. In order for this to work, an organization needs to have on the one hand an excellent understanding of its own systems and dependencies to be able to identify where configuration changes would mitigate or prevent specific cyber threats or attacks. On the other hand, it necessitates the availability of threat intelligence that provides analysis about specific attacks and discusses potential mitigations, which allows AI systems to derive appropriate automated mitigations to be applied to individual systems. Both of those conditions are fulfilled by the cybersecurity awareness concept outlined above, utilizing the threat intelligence provided by cybersecurity communities.

Cybersecurity information sharing describes the concept of sharing information of attacks or applied mitigations with the larger community. The advantage of sharing information in this context is that a community of experts can assess cyber threats or attacks based on data from many individual organizations, and derive better mitigations from this information. For example, European legislation like NIS and GDPR legally require sharing of information relating to cyber incidents under certain conditions. The main problem in this context is to be able to identify all the information that adequately forensically describes specific cyber incidents, ideally based on actual log data. An awareness system as outlined above, based on the understanding of systems and dependencies, allows to automatically identify the main information sources and extract relevant data to be shared with cybersecurity communities or authorities, which can be used to automate the compiling and submission of incident reports.

1.7 Summary and Outlook

In this Chapter we have provided an overview of the current cybersecurity landscape and developments, and which mechanisms are available on the technological, organizational and national/European/global levels to address cybersecurity problems. One of the major concerns in this respect is that cybersecurity is an extremely dynamic problem domain, which has since the early days of the global information network been a race between cybercriminals and cybersecurity experts to devise more advanced cyber-attacks and cyber defence mechanisms respectively. New legal and regulatory efforts especially in Europe acknowledge and account for this reality by taking a collaborative approach towards cybersecurity to be able to identify and react to new cyber threats and cyber-attacks more quickly in order to limit the potential impacts.

We have mapped those developments to the specific environment of LPAs, and we have seen the requirements in cybersecurity to focus heavily on protecting the data managed and processed by LPAs, including citizen and employee data with strong privacy protection requirements. The main identified concern in this regard is the lack of awareness in organizations of how data flows through and is processed by the organizational system during day-to-day operations, and what the associated implications for cybersecurity are in this context.

We have demonstrated the need for gaining awareness through a holistic understanding of the organization and its implications for cybersecurity which can only be achieved through collaboration within the organization. Furthermore, we have pointed out the need for continuous and data driven monitoring of the security state for proactive and timely cyber incident management.

In the following Chapters of this book, we will provide more insights into how the CS-AWARE project has created a cybersecurity awareness system that fills the gap identified in this Chapter in the LPA context. Chapter 2 introduces a socio-technical system and dependency analysis (SDA) methodology we applied during the project to increase the ability of LPAs to create awareness and collaborate on identifying cybersecurity relevant aspects, and which allows to identify organizational assets, dependencies, business processes, information flows and relevant monitoring points and monitoring patterns. In Chap. 3, we present a qualitative approach for capturing user awareness of cybersecurity in an organisational context. This method, a combination of storytelling and activity theoretical reasoning, adds essential aspects of the user perspective in context to the SDA methodology. Chapter 4 describes the technology stack that was implemented during the project to address to address the gap of being able to provide continuous and data driven monitoring of the security state. The system is able to process information from various sources (monitoring points identified in the SDA, and threat intelligence provided by cybersecurity communities like CERTs/CSIRTs) and correlate this information in order to detect incidents and visualize relevant context information and mitigation strategies to the user. Advanced features like self-healing and information sharing are part of this technology stack. Chapter 5 presents the deployment and validation of CS-AWARE in two LPA contexts, and shows the outcomes of both qualitative and quantitative collection of evidence from users. In particular, we show how the combination of agile software development and design-based research, applied to both usability and validation, has supported the implementation of user input in the design of the system. In Chap. 6, the Municipalities of Rome and Larissa who have been pilot partners in CS-AWARE detail their perspective of how CS-AWARE has helped to increase their cybersecurity awareness and collaboration in their organization. Chapter 7 discusses the challenges and opportunities of commercializing such a system in the European LPA marked. In Chap. 8, we present the outcomes of a study about further improving the usability of the various user options, for additional groups of users in the LPA-context. Finally, Chap. 9 reiterates the main characteristics of the CS-AWARE approach, and the preparedness of this approach for creating cybersecurity awareness in emerging and future technologies related to the LPA sector and beyond.

References

CC1. (2017). Common Criteria for Information Technology Security Evaluation Part 1: Introduction and general model April 2017 Version 3.1 Revision 5 (CCMB-2017-04-001). Retrieved from https://www.commoncriteriaportal.org/files/ccfiles/CCPART1V3.1R5.pdf

CC2. (2017). Common Criteria for Information Technology Security Evaluation Part 2: Security functional components April 2017 Version 3.1 Revision 5. Retrieved from https://www.commoncriteriaportal.org/files/ccfiles/CCPART2V3.1R5.pdf

CC3. (2017). Common Criteria for Information Technology Security Evaluation Part 3: Security assurance components April 2017 Version 3.1 Revision 5 (CCMB-2017-04-003). Retrieved from https://www.commoncriteriaportal.org/files/ccfiles/CCPART3V3.1R5.pdf

Commission of the European Communities. (2006). COMMUNICATION FROM THE COMMISSION on a European Programme for Critical Infrastructure Protection (COM(2006) 786 final).

Coppolino, L., D'Antonio, S., Mazzeo, G., Romano, L., & Sgaglione, L. (2018). How to protect public administration from cybersecurity threats: The COMPACT project. In *2018 32nd International Conference on Advanced Information Networking and Applications Workshops (WAINA)* (pp. 573–578). https://doi.org/10.1109/WAINA.2018.00147

ENISA. (2019). ENISA threat landscape report 2018: 15 top cyberthreats and trends. https://www.enisa.europa.eu/publications/enisa-threat-landscape-report-2018/at_download/fullReport

ENISA. (2020). Cybersecurity certification. EUCC, a candidate cybersecurity certification scheme to serve as a successor to the existing SOG-IS (No. V1.0, 01/07/2020; p. 283). European Union Agency for Cybersecurity.

European Commission. (2013). Proposal for a DIRECTIVE OF THE EUROPEAN PARLIAMENT AND OF THE COUNCIL concerning measures to ensure a high common level of network and information security across the Union (Brussels, 7.2.2013 COM(2013) 48 final 2013/0027

(COD)). European Commission. Retrieved from https://edps.europa.eu/data-protection/our-work/publications/opinions/cyber-security-strategy-european-union-open-safe-and_en

European Commission. (2020a). Proposal for a DIRECTIVE OF THE EUROPEAN PARLIAMENT AND OF THE COUNCIL on measures for a high common level of cybersecurity across the Union, repealing Directive (EU) 2016/1148 (COM(2020) 823 final, 2020/0359 (COD)).

European Commission. (2020b). Proposal for a DIRECTIVE OF THE EUROPEAN PARLIAMENT AND OF THE COUNCIL on THE resilience of critical entities (2020/0365 (COD)).

European Parliament. (2016a). Directive (EU) 2016/1148 of the European Parliament and of the Council of 6 July 2016 concerning measures for a high common level of security of network and information systems across the Union. Official Journal of the European Union, L194/1.

European Parliament. (2016b). Regulation (EU) 2016/679 of the European Parliament and of the Council of 27 April 2016 on the protection of natural persons with regard to the processing of personal data and on the free movement of such data, and repealing Directive 95/46/EC (General Data Protection Regulation). Official Journal of the European Union, L119/1. https://eur-lex.europa.eu/legal-content/EN/TXT/PDF/?uri=CELEX:32016R0679&from=EN

European Parliament. (2018). Directive (EU) 2018/1972 of the European Parliament and of the Council of 11 December 2018 establishing the European Electronic Communications Code (Recast)Text with EEA relevance. *Official Journal of the European Union, L321*(36), 179.

European Parliament. (2019). Regulation 2019/881 of the European Parliament and of the Council on ENISA and on information and communication technology cybersecurity certification. Official journal of the European Union, L151/15.

Europol. (2019). Internet organised crime threat assessment 2019. European Union Agency for Law Enforcement Cooperation.

Europol. (2021). *Internet organised crime threat assessment 2021*. European Union Agency for Law Enforcement Cooperation.

GovCERT.ch. (2016). APT case RUAG technical report. GovCERT.ch. https://www.govcert.ch/downloads/whitepapers/Report_Ruag-Espionage-Case.pdf

Hsiao, S.-C., & Kao, D.-Y. (2018). The static analysis of WannaCry ransomware. *International Conference on Advanced Communications Technology (ICACT)*, 153–158.

Hybrid CoE. (2021). Hybrid CoE's key themes and approaches to countering hybrid threats in 2021. The European Centre of Excellence for Countering Hybrid Threats. https://www.hybrid-coe.fi/wp-content/uploads/2021/04/Workplan2021_summary_rgb.pdf.

ISO. (2013). ISO/IEC 27001:2013 information technology—Security techniques—Information security management systems—Requirements. ISO. https://www.iso.org/cms/render/live/en/sites/isoorg/contents/data/standard/05/45/54534.html

Lamba, A., Singh, S., Singh, B., Dutta, N., & Muni, S. S. R. (2017). Analyzing and fixing cyber security threats for supply chain management. *SSRN Electronic Journal*. https://doi.org/10.2139/ssrn.3492687

NIST. (2018). Framework for improving Critical Infrastructure Cybersecurity, Version 1.1 (NIST CSWP 04162018; p. NIST CSWP 04162018). National Institute of Standards and Technology. https://doi.org/10.6028/NIST.CSWP.04162018

Touhill, G. J., & Touhill, C. J. (2014). *Cybersecurity for executives: A practical guide*. John Wiley & Sons.

Chapter 2
The Socio-Technical Approach to Cybersecurity Awareness

Christopher Wills

2.1 Introduction

In this Chapter we introduce how CS-AWARE implements one of its fundamental principles; the supposition that an effective cybersecurity management in an organisation—including the technology that supports the cybersecurity management requires a holistic awareness and understanding of the socio-technical system set-up that influences the cybersecurity of the organisation, and the interactions between those systems. We show how this awareness can be reached through collaboration within the organisation, and how we can utilize the gained holistic understanding as a basis for the technological part of the CS-AWARE cybersecurity awareness solution.

In order best to understand the approach taken in the CS-AWARE project to the systems dependency analysis (SDA) in the two pilot cities and the subsequent design of each pilot city's CS-AWARE system, it's necessary to have some insight into the socio-technical approach that was adopted and applied in the project. This socio-technical approach was used in a holistic analysis of both an organisation's cybersecurity, and of its information systems architecture. In Sect. 2.2 we first introduce the concept of socio-technical thinking in general and more concrete approaches for soft systems analysis, like the Soft Systems Methodology (SSM) by Peter Checkland that was utilised in the CS-WARE project. Section 2.3 describes how the approach was adapted for the CS-AWARE project to achieve the desired level of awareness and collaboration within the organisation, and to achieve the desired depth of analysis required to serve as input for the CS-AWARE awareness monitoring technology,

C. Wills (✉)
Caris Partnership, Fowey, UK
e-mail: chris.wills@cs-aware.com

© The Author(s), under exclusive license to Springer Nature Switzerland AG 2022
J. Andriessen et al. (eds.), *Cybersecurity Awareness*, Advances in Information
Security 88, https://doi.org/10.1007/978-3-031-04227-0_2

including a practical step-by-step guide of how to conduct the analysis workshop sessions. Section 2.4 concludes the chapter and discusses the results.

2.2 Introducing the Concept of Socio-Technical Systems

The term "Socio-technical System" has its roots in Systems Theory. Systems Theory and Systems Thinking are simply a way of looking at some part of the world, by choosing to regard it as a system, using a framework of perspectives to understand its complexity and to undertake some process of change. The key concepts are holism—looking at things as a whole and not as isolated components and systemic—treating things as systems, using systems ideas and adopting a systems perspective.

The term "System" describes an organised entity that incorporates and connects all its components (be they biological, mechanical, digital or human). When operating correctly, a system processes inputs and generates an expected output in a predictable manner.

As depicted in Fig. 2.1, systems can be divided into two broad types; "hard" systems and "soft" systems. Hard systems and hard systems thinking, is founded on mathematically-based systems analysis and systems engineering. It assumes that the world is comprised of systems that we can describe accurately and that these systems can be understood through rational analysis. It is based on the assumption that it is possible to identify a "technically optimal" engineering solution for any system. It assumes that we can then write software to create the "solution".

Hard systems thinking, assumes that there is a clear consensus as to the nature of the problem that is to be solved. However, it is unable to depict, understand, or make provisions for, unquantifiable variables, such as people, culture, politics and aesthetics. It also assumes that those commissioning the system have the ability and power to implement the system.

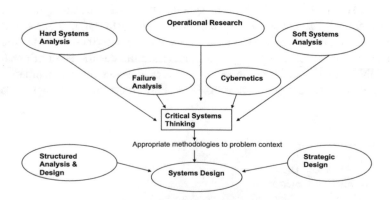

Fig. 2.1 Systems thinking—The systems approach

The hard systems design approach is very good for engineering physical or systems structures that require little or no continuing inputs from people, such as the design and construction of a suspension bridge, or automated process control in an oil refinery, or for control systems for robots used in manufacturing processes, etc. However, for the reasons we have set out above, the hard systems approach cannot analyse, depict, describe or design systems that rely upon inputs from, or interactions with, people.

Systems that require human interaction are often referred to as "human activity systems" or "socio-technical systems". Socio-technical systems are those systems whose operations, processes and outputs require some sort of sophisticated and continuing human intervention and interaction in order for the system to work. Any system that requires the input, manipulation or interpretation of data or information by a human is, by definition, a socio-technical system.

2.2.1 Socio-Technical Systems and Soft Systems Design

The dawning of our understanding of Socio-Technical Systems (STS) is usually cited as emerging from the work of Eric Trist and Ken Bamforth (Trist & Bamforth, 1951). Their study examined the mechanization of coal mining. The introduction of mechanization in some British mines had led to a lower-than-expected increase in the amount of coal being mined. Prior to mechanization, the miners worked in three shifts with each shift containing a multidisciplinary team of men with a combined skill-set which enabled them to overcome difficulties encountered in mining the coal during their shift. The first shift cut the coal, the second loaded it for transport to the surface, the third moved the equipment forward to the new coal-face in preparation for the first shift to return and cut the coal. However, each shift contained men with specialist skills which combined and covered all aspects of mining.

When mechanization was introduced, these multidisciplinary teams were disbanded and replaced with teams specialised and skilled only in the particular function of their shift—cutting, clearing away, resetting the machinery at the face. When a shift encountered a problem that was outside their specialism, they had to wait for the return of the shift that contained the appropriate skill-set, who were then able to find a solution to the problem at hand, so that mining could then recommence.

In short, the new system had been designed in such a way as to optimise the mechanization of the coal mining system. However, the human dimension of the process had been overlooked. What Trist and Bamforth's study proved was that systems that relied upon human intervention were far greater than the sum of their mechanical/ technological parts. They could only be properly understood by reference to both their technical and human components. They were not simply "technological" systems, but "human activity" systems; they were "socio-technical systems" because they were sum of both their human (socio) and technological parts. The design of such systems therefore needs to encompass and address both the social and the technological components, hence the term "Socio-Technical Design" coined by Enid Mumford (Mumford, 2003). Building on the work of Trist and Bamforth, Mumford developed what she called Socio-Technical Design. In her approach, both the technical and social requirements and components of a system were equally important. Neither the technical, nor the social components of a "human activity system", should be optimised to the point of detriment to either.

Mumford and Henshall (1983) argued that, "The training given to systems analysts is, to say the least, very much biased towards computer systems design, data manipulation and organisation techniques. It recognises the human element of the system in an almost apologetic and certainly mechanistic fashion.... This is not to say that systems analysts do not think about the human factor; they do, but they do not have the methods, tools and training at their disposal which would allow them to design systems that satisfy both the technical system and the social system requisites."

The authors are of the opinion that it is still too often the case that "human factors" though not ignored, are not fully and adequately addressed in the design of many current IT and cybersecurity systems.

Mumford (1983) identifies three levels of involvement, or user participation. They are, in ascending order of the extent of user involvement in the design process: *consensus participation*, where all the users affected by the system are included in membership of the design group; *representative participation*, where only representatives of the affected users are part of the design team; and *consultative participation*, where the users or their representatives are consulted at various stages of the design process, but not closely involved in participating in the design of the system.

The extent to which the users can participate in the design process is determined to some extent by the willingness of the host organisation to devolve the responsibility for design. To a greater extent, it is determined by the number of users affected by the system. It is more difficult to extend direct (consensus) participation in the design process to a large group of users than it is to a small group.

Direct participation by a large group in unwieldy and impractical, the logistics involved in a large group of people meeting regularly in a decision-making forum are complex and costly. Representative participation in the design of the system is also potentially problematic, for much the same reasons that representative political democracy can be problematic. Consultative participation may not be participative enough to involve the users to the point where they will feel that they own both the process, and the system that results from it. While it is clear that although the participative approach to design is not without problems it is also clear however, even at the consultative level, it still engenders a feeling of ownership of the system amongst the users.

Engendering a feeling of ownership is important. Participation by users in the process of change results in them feeling less threatened by change. A reduced perception of threat is likely to reduce the users' resistance to change. As important, is the deep-rooted emotional feeling of ownership all of us experience when we embody our intellect and creativity in the creation of something external to us. Users who feel that they have a personal stake in the system will be highly supportive of it.

The other outcome of user participation is that the knowledge elicitation process is made much more effective: "Firstly, it is a fact that the people with the greatest knowledge of the existing formal work system, and the people with the most knowledge of the informal system, are the people who are responsible for their operation. They therefore have the greatest potential to successfully design a system which overcomes the present systems shortcomings" (Mumford & Henshall, 1983, pp. 120).

Notwithstanding these practical justifications for using an STS approach for the design of information systems, there is also a moral philosophical dimension to the argument. People should be involved in the design of the systems with which they work. They should work in situations where they rewarded equally well in both intrinsic and extrinsic terms.

Mumford's STS methodology as depicted in Fig. 2.2, ETHICS (Effective Technical and Human Implementation of Computer-based systems), has 14 stages, the core method involves the following:

- Forming the design group, be it on a consensual, representative, or consultative basis.
- Appointing an "expert" analyst designer, who is attached to the group, to act in the role of "facilitator", helping the group reach conclusions about what is feasible, by offering expert advice, rather than by directing the group to a particular conclusion.
- The design group then identifies the primary social and technical requirements of the future system. The group creates a list of social alternatives, and a list of

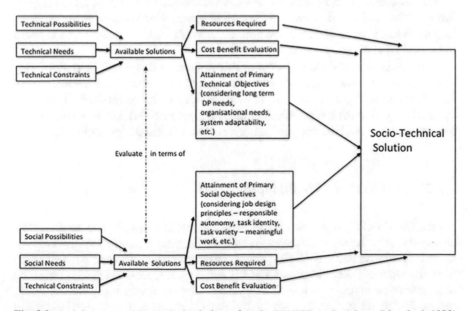

Fig. 2.2 Arriving at a socio-technical solution using the ETHICS methodology (Mumford, 1993)

technical alternatives, which will contribute to attaining the primary social and technical objectives. These two lists will contain those features that should be embodied in the future system, in rank order of desirability.

- Social alternatives will focus on those features that provide the best social solution to the problem, (typically, those features which improve the quality of working life of those who will work with the system, or are affected in some way by it). These features will be considered in terms of what is possible (social possibilities), what is desirable (social needs), and what is undesirable or prohibitive (social constraints).
- Technical alternatives will be concerned with those features which will provide the greatest technical improvement to the problem, and as with social alternatives, will be considered in terms of what is possible (technical possibilities), what is desirable or prohibitive (technical needs) and (technical constraints).
- Comparisons are then made between the social and technical alternatives, and a list containing those social and technical alternatives, which are not mutually exclusive, is compiled.
- These potentially available solutions are now evaluated in terms of costs/benefits, and the extent to which they satisfy both the primary social and technical objectives, the outcome of this process being the highest-ranking STS solution.

Latterly, Mumford's work has been used as a base for other STS approaches (HUSAT, 1988), and latterly has been incorporated with other approaches into a flexible methodology called Multiview (Wood-Harper et al., 1985), detailed discussion of which is beyond the scope of this chapter. Moreover, Mumford's participative approach is also reflected in current "AGILE" development methods.

The approach taken in the CS-AWARE workshops (in the pilot cities) largely follows Mumford's lead in terms of the design group, which in both of the pilot cities, involved the participation of the greatest possible number of stakeholders, both in the SSM and the story telling workshops.

Cybersecurity is a socio-technical problem because it is utterly and absolutely reliant upon the knowledge, appropriate behaviour and timely intervention of personnel, in order for data and information security to be maintained. Therefore, cybersecurity systems are usually (although not exclusively), socio-technical systems and as such are best designed using a socio-technical design approach.

2.2.2 Soft Systems Methodology

For the CS-AWARE project we have utilised the Soft Systems Methodology (SSM), an approach that builds upon the socio-technical approach and soft systems design to analyse existing and organically grown complex systems. The approach emerged from the work of Peter Checkland and his colleagues. Checkland (1981) observed that systems involving people—"Soft Systems" are often very complicated, fuzzy, messy, ill-defined and are typified by unclear situations, differing viewpoints and unclear objectives, which include politics, emotion and social drama. This is the type of system domain for which Checkland's Soft Systems Design approach, is highly appropriate and to which it should be applied. This is of course the case with the analysis and design of a cybersecurity awareness system such as was the focus of the CS-AWARE project.

The idea underpinning SSM is that of creating a comparison between what can be seen in the real world and what can be examined using systems thinking, as depicted in Fig. 2.3.

Checkland's Soft Systems Methodology has seven steps, as seen in Fig. 2.4, which form the approach. Although the methodology involves seven stages, it may not be necessary to apply all seven to a given problem situation. In some cases, interventions that significantly improve the problem domain become extant in the earlier stages of the method. The first stage is that of exploring and thinking about the problem situation that is under consideration. As is set out above "Human Activity Systems" are often messy and ill-defined, contain sometimes few, if any, measurable objectives and consist of differing views coloured by organisational politics typified by high social drama. The second stage involves creating "Rich Pictures" (RP's).

RP's are a representation of the problem domain. They utilize "cartoon-style" techniques to portray a complex situation and concentrate on:

- Structure—Key individuals, organisations etc.
- Process—What could be or is happening
- Climate—Pressures, attitudes, cultures, threats etc.

RP's are a tool for understanding a problem situation whether it's a system or not. They are a mixture of drawings, pictures, symbols and text. They represent a particular situation or issue and they are created from viewpoint(s) of the person or people who drew them. They can both record and evoke insight into a situation.

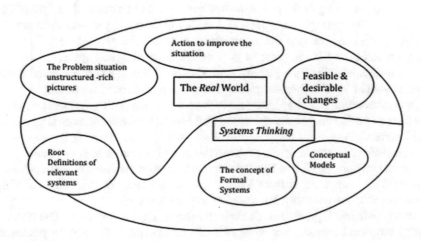

Fig. 2.3 SSM core concepts

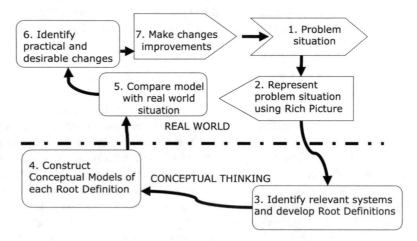

Fig. 2.4 The seven steps of the SSM methodology

RP's are pictorial 'summaries' of a situation, embracing both the physical, conceptual and emotional aspects of a problem situation. There are no rules that apply to the drawing of Rich Pictures. There is no right or wrong way to construct an RP, it is entirely in the hands of the person, or the group of people, who are drawing the picture. No artistic skill is required!

The process of creating a RP is simple. The stakeholders taking part in the SDA, construct a diagrammatic representation of the problem domain. This RP depicts both the physical (logical) and the political social and cultural elements that they have identified in the problem situation that is being examined. The full participation of stakeholders at this stage of the method is essential, as they will have a far deeper understanding of the problem situation than will an external analyst or consultant. The RP attempts to capture and represent both the physical infrastructure and the processes, but also the relationships between the stakeholders. These are represented in the RP's in order to gain an understanding and appreciation of the problem situation. The Rich Pictures should contain 'hard' information and factual data, as well as 'soft' information; subjective interpretations of situations, including aspects of conflict, emotions etc. The RP should give a holistic impression which can then be further analysed, refined and understood by the subsequent stages of the SSM methodology. RP's are powerful; they enable those drawing the pictures to express their understanding of a problem domain or a system in great detail.

As an example, the RP in Fig. 2.5 depicts the respective check-in operations of two mythical competing airlines, Pacific Air and Atlantic Air. This RP is machine drawn for the sake of clarity. RP's are usually hand drawn.

Pacific Air has a problem with their passenger check-in system. The check-in clerk's terminal keeps malfunctioning and the clerk cannot check-in the passengers and their luggage. The customers are understandably disgruntled, as are the

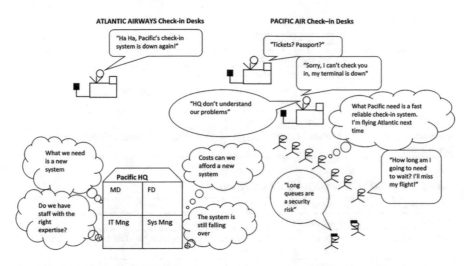

Fig. 2.5 Example of a Rich Picture

check-in staff. Pacific Air's management have a variety of views from the Managing Director understanding that they need a new system, the Financial Director is worried about where money will come from to pay for a new system, the IT manager is worried about whether they have the right staff with the right expertise and is concerned that that's perhaps why the system keeps failing. The airport management are concerned about the security risk that the long queues at check-in creates, as none of the passengers have passed through security at that point. Atlantic Air meanwhile is delighted, because they hope to pick up more passengers as a result of Pacific Air's failure.

Once the Rich Picture has been constructed, in third stage, relevant sub-systems are identified and defined using "Root Definitions". A Root Definition is not an attempt to describe some real, existing 'system'. It is an attempt to learn about a complex situation to enable changes to be made which encapsulate the essence of these systems, the 'whats' (what does it do?) rather than the 'hows' (how does it do it?). The Root Definitions describe the core transformation activities and processes of the system—the conversion of inputs into outputs. A Root Definition of the RP above would be:

> An airline owned passenger check-in system that enables passengers to check in their baggage and enables airline staff to issue passengers with boarding cards in a manner that is consistent with the safe and timely operation of the airline's departure schedules.

Checkland used a mnemonic—"CATWOE" to check that all of the necessary components of a problem situation were represented in any a RP of that situation so as to ensure that the 'Root Definitions are complete and accurately represent the problem situation portrayed in the Rich Picture (Table 2.1).

A "CATWOE" of the airline RP would be as follows (Table 2.2):

Table 2.1 CATWOE

Customers	Those who benefit in some form from the system
Actors	The people involved
Transformation	The development of outputs from inputs
Weltanschauung	The 'world view' a holistic overview of both the transformation processes and the problem situation
Owner	The person(s) with control
Environmental constraints	Physical boundaries, political, economic, ethical or legal issues

Table 2.2 A "CATWOE" of the airline Rich Picture

Customers	The passengers
Actors	The airline staff
Transformation	Unchecked baggage become checked, passenger tickets supplemented with boarding cards
Weltanschauung	The efficacious effective and efficient operation of the airport and the airline (it works with minimum waste and meets the expectations of the passengers, the airline and the airport)
Owner	The airline
Environmental constraint	Time, safety, security effectiveness (passengers and baggage departing to the same correct destination)

The fourth stage of the method is that of creating conceptual models. Once the Root Definition(s) have been constructed and have been compared with the Rich Picture and checked against CATWOE, Conceptual Models can be constructed. The Conceptual Models are formed from the actions stated or implied in the Root Definition(s). Of course, each Rich Picture may be interpreted from quite differing 'world view points' Conceptual Models may be derived from a Root Definition even though knowledge of any 'real-world' version of the activity is lacking.

A Conceptual Model is like an activity sequence diagram, but is aimed at representing a conceptual system as defined by the logic of the Root Definition and not just a set of activities. It is not a representation of what exists in reality, nor is it necessarily a representation of what ought to exist. It should contain only those actions that would have to be carried out, and the order in which they would have to be carried out, if the system in the Root Definition were to function. Conceptual models will differ depending on the of the view-point of the observer. The viewpoints and understanding of the airline RP by the check-in staff, the Financial Director and the IT Director, are all quite different.

Having developed a top level, primary activity Conceptual Model as the one illustrated in Fig. 2.6, each of the activities identified are modelled in a second level Conceptual Model like the one shown in Fig. 2.7, once again based on our airline problem situation.

In the fifth stage the Conceptual Model(s) are checked against both the Root Definition(s) and the Rich Picture. A good way of performing these checks is to ask three questions: Do the activities exist? Who does them? Why do it that way?

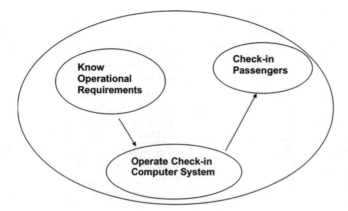

Fig. 2.6 A Top-level Conceptual Model of the Check-in Process

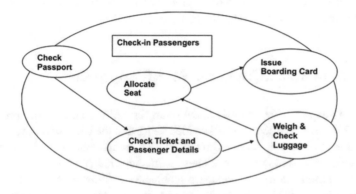

Fig. 2.7 A Second Level, Secondary Activity Conceptual Model

The analyst(s) need to imagine that the Conceptual Model is actually operating in the real world. They can identify a real process from the Rich Picture, follow its sequence in the Conceptual Model and compare how the sequence would operate in reality. This process can be represented using a chart (Table 2.3):

The sixth stage of the methodology is that of identifying practical interventions (changes) that can be introduced into the problem domain to improve the current situation. Usually, the interventions that are required are obvious and emerge as a result of drilling down through successive layers of conceptual models. Sooner or later, practical and practicable interventions (changes) that will improve the problem situation become apparent. There may be several interventions that can be made. These may be social, technical organisational or economic in nature. In the preceding example of Pacific Air's check-in system, is it the case that system too old, and has become obsolete and difficult to maintain, as the Managing Director suspects? Perhaps the skill set of the technical staff needs to be improved, as the IT manager believes? Perhaps some coding errors have somehow been introduced into

Table 2.3 Real world processes

Activity in conceptual model	Present in real world situation (rich picture)	Comments	Include on Agenda
Check passport	Yes	Process tasks place independent of check-in computer system	No
Check passenger and ticket details	Yes	Dependent upon operation of check-in computer system	Yes
Weight and check luggage	Yes	Process tasks place independent of check-in computer system	No
Issue boarding card	Yes	Dependent upon operation of check-in computer system	Yes
Allocate seat	Yes	Dependent upon operation of check-in computer system	Yes

the booking system software? Using SSM to drill down through all of these socio-technical aspects of the problem domain will reveal the issues and the appropriate and practicable solutions represented by the methodologies seventh and final stage.

2.3   SSM in Action: The CS-AWARE Approach

The systems dependency analysis (SDA) that formed the core of the project was based on SSM as described above. The idea being that the users of an organisation's systems use SSM to help them express and give an account of, their (often tacit), knowledge in dedicated workshops. These workshops included participants from all levels of the organisation (e.g. managers, technicians, administrators and end-users).

The goal of the SDA is to be able to interface an organisation's system to the CS-AWARE technical solution for continuous cybersecurity monitoring and awareness. For this purpose, three main aspects need to be derived:

- Assets, dependencies and monitoring points: Identify critical assets, the dependencies between those assets and potential monitoring points (log files) to surveil those assets.
- Business processes and information flows: Identify critical business processes for the organisation's day-to-day operations, and the information flows that those processes produce through the organisation's systems (assets and dependencies).
- Observable parameters and system behaviour: Identify how the system behaviour (and the boundaries between normal/abnormal behaviour) is reflected within the monitoring points (log files) and identify observable parameters in those files that capture this be-haviour.

The following Sections will describe the procedure that was followed to conduct the analysis workshops as a step-by-step guideline. This guideline was originally published in CS-AWARE deliverable D2.5 "Guidelines and procedures for system and dependency analysis in the context of local public administrations". The deliverable should be consulted for further detail and context, for example concerning suggestions for how to organize SSM workshops in an organisation.

2.3.1   A Practical Guide to the CS-AWARE Analysis Approach

This Section describes in detail the three stages of SSM analysis conducted in the project. The first stage is concerned with structural analysis of assets and dependencies, the second stage is concerned with analysis of business processes and information flows, and the third stage is concerned with behavioural analysis and identification of monitoring patterns.

2.3.1.1   Structural Analysis: Understanding the Socio-Technical System in Terms of Assets and Dependencies

The context of this stage of the analysis is to gain a holistic picture of the critical assets and their dependencies within an organisation that are relevant for cybersecurity monitoring and awareness. These include (but are not limited to), the perspective of the user that interacts with those systems, the manager that oversees the systems operations, and the admins/technicians that ensure day-to-day operation of the systems. The concrete set-up of each organisation will be highly dependent on each organisation, but the following procedure should guide through a successful workshop:

1. Start the workshop by giving a general overview of the context and purpose and the expected outcomes and results of the workshop. All participants should have received this guide before for pre-workshop preparation, but a general introductory session will ensure a common understanding. Reserve time for comments and questions.
2. Ask the workshop participants to identify the most critical assets and/or assets related to the most critical business processes of the organisation. While those will be highly dependent on the organisational context, a good starting point are assets related to:

   (a) Financial systems of the organisation
   (b) Systems that relate to data that is managed by the organisation, for both public sector and commercial organisations
   (c) Systems that relate to services provided to a large customer base (citizen services in the public sector, commercial products in the private sector)

Depending on the size of the group, divide the workshop into groups no larger than five for providing an initial list of critical assets. The groups should prepare their results in form of a Rich Picture and present it to the workshop participants. It is to be expected that the input and discussions in the larger group will provide additional viewpoints that substantiate the initial results. If necessary, repeat this step by shuffling the group participants until a consensus among the participants about the most critical assets is found.

3. At the next stage, the goal is to concentrate on the initial list of critical assets, and identify all the relevant elements and dependencies that are required by those assets to operate, from the interactions on the end user side, to the technical elements that process and store data. While the concrete set-up highly depends on the organisational set-up, experience has shown the elements required to ensure service operation are usually part of:

   (a) the network infrastructure,
   (b) the application service infrastructure,
   (c) databases, and
   (d) security appliances.

4. Participants should work in groups no larger than five to create a high-level picture of their understanding of the organisation's assets, dependencies and interactions between the two. Groups should be formed that are expected to contain perspectives from different organisational levels. For large and complex organisational set-ups, it may be necessary in the first round, for the groups to work on different aspects or the systems (e.g. network, services, security, ...). For low to medium complexity set-ups, each group can aim to develop a holistic understanding in the early phases.

5. The results of the group discussion should be put to a Rich Picture on a flip chart, after which each group should present the results of their discussion to the other groups. At this stage it is to be expected that a lively discussion among the workshop participants will take place. This will reveal the tacit knowledge and differing view-points of participants, and will facilitate the development of a holistic understanding of the system. After this exercise, the groups should be asked to develop and refine their understanding, and provide a more detailed overview, including the input from the first round of discussions. If necessary, re-order the groups based on outcomes of first round. Results should again produce a Rich Picture, which is presented to the analysts and other participants. This step should be repeated until a consensus among the workshop participants on the holistic understanding of the organisations systems and dependencies is agreed.

6. As a last step for information collection related to the first stage of system and dependency analysis, potential monitoring points, data sources (log files) that allow continuous monitoring of previously specified assets, need to be identified. As in previous steps, it is to be expected that those sources will relate to elements on the network, service, database and security appliance level. Based on the previously achieved common understanding, it may be possible to discuss this aspect in the large group. However, it may be necessary to follow a similar process as before and divide the workshop into groups, each depicting the results on Rich Pictures and presenting the results of each group to all of the workshop participants.

Once a consensus on the holistic understanding of the organisation's asset and dependency set-up has been achieved and potential monitoring points are identified, and transferred into an asset and dependency graph, the first stage of system and dependency analysis is concluded. It is important that, once the workshop results are analysed, an additional workshop session including all participants is required to double-check that the analysis is correct, complete and represented in the graph and amend any missing information. This can be done in preparation for stage 2 of the workshop, or in the context of the first workshop session of stage 2.

2.3.1.2   Business Process and Information Flow Analysis

While the objective of the first stage of the analysis is to understand the system in terms of assets and dependencies, the goal of this stage is to identify how day-to-day operations map to the asset and dependency graph in terms of information flows caused by relevant business processes. The following procedure should guide the reader through a successful workshop:

1. Start by reviewing the system/dependency graph from the first stage. If this has not been done in between the workshops, reserve time for reviewing the system and dependency graph and allow for modifications until a consensus is reached among the participants.
2. Based on the critical systems/services identified in the first stage, give the participants some time to define business processes that are required to conduct day-to-day operations within those services. If deemed necessary, the participants can be asked to prepare for this task before the workshop. Depending on the number of participants, groups may be formed, with each group focusing on one of the identified critical systems/services. Make sure that multiple organisational viewpoints are present in each group.
3. The resulting list of business processes should be visualized in the form of a Rich Picture, and each group should present their results to the workshop participants. It is to be expected that the understanding of business processes will be complemented by the input/discussions within the group. If necessary, additional rounds of group work, followed by presentations of results, should be conducted until a consensus is reached. The membership of groups may be shuffled if necessary.
4. Once a consensus is reached, the workshop can move on to define the business processes in detail. Experience has shown that the CATWOE method described in Sect. 2.2.2 is an excellent approach to analyse business processes.
5. For each process identified, a CATWOE list should be produced. Depending on group size and the number of processes to analyse, this can be either done with all workshop participants present, or in group work. The results should be presented as a Rich Picture and presented to the group. The process should be repeated until a consensus is reached for each business process CATWOE list.
6. As a result of the CATWOE analysis, the identification of the actors and data transformations provides sufficient information to map the business process to

the asset and dependency graph, by identifying the information flows resulting from the processes identified through the organisation's systems. For each process, identify which assets are involved in conducting the process, from end-user interaction, to communication over network and from data processing by the back-end service to data storage. It may turn out that relevant assets were previously omitted, which should be complemented accordingly.
7. A review of potential monitoring sources (log files) to observe system behaviour should be conducted, and it should be identified if additional monitoring sources come to light in the context of business process observation.

Once a consensus regarding the information flows per business process has been achieved, the second step of SDA analysis is concluded. After the information flows are modelled within the CS-AWARE asset and dependency graph, they should be discussed with the workshop participants. This can be done in preparation for, or during the workshop sessions of the third stage.

2.3.1.3   System Behaviour Analysis

Having derived structural information in stage 1 and process information in stage 2, the only missing aspect to interface an organisation's system to CS-AWARE for continuous cybersecurity monitoring is that of the definition of behaviour that day-to-day operations relating to the identified business processes, and how this behaviour is reflected in the monitoring points (log files) identified in steps 1 and 2. Moreover, the boundaries that determine normal and abnormal behaviour as recorded by various parameters logged within the log files, are the basis for the definition of monitoring patterns by CS-AWARE experts. In general, there are two main pattern types relevant for CS-AWARE: (1) behaviour-based monitoring patterns that monitor behavioural patterns in order to detect abnormal and potentially malicious behaviour, and (2) indicator-based monitoring patterns that associate uniquely identifiable security information (vulnerability, virus, network intrusion, …) in order to associate context for awareness.

Mapping abstract behaviour to concrete and, in most cases, deep technical information found in log files is a more involved process than in the two previous phases—especially for workshop participants that do not usually think in technical terms. Experience has shown however that with careful preparation and by following the process laid out in this guide, the contributions from workshop participants will lead to significant results with regards to identifying abnormal or suspicious system behaviour and to how this behaviour is reflected in log data. For this to be successful, careful preparation by the analysts before the workshop is required: A sample of all log files used as monitoring sources, as identified in stages 1 and 2, should be investigated and all the logged parameters within those files should be identified. Monitoring sources are of course highly dependent on the organisational set-up and will require individual analysis. A few generalizations can be given regarding the expected data on the database, the service, the network and the security appliance level. Both databases and application services usually keep audit logs for data operations and authentication, which are highly relevant for behaviour monitoring. The network and firewall security appliance logs usually contain various aspects relating to network traffic, and are both relevant for behaviour and indicator-based monitoring. Antivirus and IDS/IPS security appliances logs detected events and are usually relevant for indicator-based monitoring. Once pre-workshop preparation is completed, the following procedure should guide participants through a successful workshop:

1. When the workshop starts, a quick recap of the results of the first stage (system and dependency graph) and second stage (business processes/information flows) should be presented. Allow for time to capture additional input from workshop participants for corrections/ additions that have come to light since the previous workshop sessions. If necessary, repeat relevant analysis elements from stages 1 and 2.

2. The previously identified monitoring points (log files) should be introduced. Based on the pre-workshop preparation, a general over-view of the log file structure and the available parameters should be given. Together with the participants, the meaning of each parameter in the specific context of the organisation and relevant business process should be defined in discussion until a consensus is found. While many parameters have a clear purpose and meaning, there may be parameters that are context specific and require input from the organisational context to be fully understood.

3. The review of the log files establishes a common understanding of the log file contents, and the information about system behaviour that is captured by each parameter. The next step is to ask workshop participants to define behaviour and/ or scenarios that

    (a) Are considered disruptive or malicious to the various business processes reflected in the data,
    (b) Are something that the organisation (represented by the roles of managers, technicians, service users, …) want to be made aware of,
    (c) Are something that is not part of the current monitoring activities and/or something that cannot easily be monitored within the current set-up.

    Depending on the group size, the workshop can be divided into groups, while making sure that different organisational perspectives are present in each group. The results of the group discussions should be presented to all workshop participants, discussing and substantiating the scenarios by input from other group members. Those scenarios that are deemed relevant by the consensus of the group, are to be further investigated in the next steps. Unrealistic or irrelevant scenarios should be discarded by consensus. This exercise may be repeated, shuffling the group participants, until no more relevant scenarios can be identified.

4. Identify parameters in the log files that are able to capture behaviour of the scenarios defined in the previous step and how this behaviour would be reflected in

the data logged by those parameters. For behaviour/scenarios, it is expected that the most relevant log files will be on the database and service level.

5. For each identified behaviour reflected in one or multiple parameters, the boundaries between normal and abnormal behaviour are to be defined by consensus of the workshop participants. For example, if a behaviour/scenario to be monitored is the access time to a specific service (logged by the authentication log), the normal behaviour may be expected as login during business hours (e.g. 9:00 to 17:00). However, the tacit knowledge of workshop participants may reveal there are users that work late, which would extend the normal operating hours (e.g. 9:00 to 20:00). Repeat above steps until all identified scenarios have been discussed and a consent was found for parameters that capture the specific behaviours and their boundaries between normal and abnormal.

6. Especially the log sources that are mainly relevant for indicator-based monitoring (e.g. security appliance logs) may not be involved in any of the discussed behavioural scenarios. While they may not require input from the workshop participants in order to derive unique event identifiers that can be utilised by CS-AWARE for providing additional context, they should still be discussed during the workshop sessions, since participants may have additional tacit knowledge about the organisational context of the log sources that may help to derive more relevant monitoring patterns.

2.3.2   Examples from the CS-AWARE Pilot Use Cases

Several workshops conforming to the procedures described in the previous sections were run in each of the pilot cities. At the start of the series of workshops in both of the cities, an overview was given of the Soft Systems Methodology. The participants in each workshop were asked to draw a RP of the mission-critical systems in their respective cities. As we have previously set out, and as we discuss further below, we defined mission-critical systems as being those systems which contained or processed either sensitive personal information (such as would fall under the provisions of the GDPR), or information or processes which were vital for the city's ability to provide vital services to its citizens. This would include systems and processes that were used to maintain and handle the city's finances.

Figure 2.8 show examples of such a "top level" RP's in Larissa and an associated description of the applications, systems and networks represented in each of the RP's.

Larissa RP 1 above gives an overview of the city's network and main services. The only gateway from the City's systems to the Internet and telephony to the outside world is via a router provided by SYZEFXIS. Syzefxis is an entity that runs under the ministry of digital policy, and it provides interconnection between all the public sectors in Greece. Syzefxis also provides the main network gateways to each public entity. The, so called, SYZEFXIS router is run by the Syzefxis team. The Cities of Larissa, Elassona and Kileler all use the SYZEFXIS router. SYZEFXIS is

Fig. 2.8 Top Level Rich
Picture Larissa

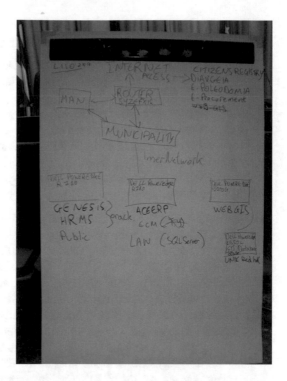

connected to both the Metropolitan Area Network (MAN) and to the three servers located in the Town Hall. The R710 host Genesis, (the City's ERP system that is used to manage income and expenditure). Genesis handles information about both suppliers and citizens and tax and debt collection. The system is used for maintaining the civil registry, for records of document signing and for the cash desk.

In Rome, a similar set of high-level Rich Pictures was produced, which are not included in this chapter due to their sensitive nature. Rome's data are held both inside and outside the city. In Milan there are two data centres in two different locations in the city of Milan working an active-active cluster, accordingly to a Business Continuity architecture. Data centres in Milan are owned by a supplier, that also provides Roma Capitale with Internet and network perimeter security services. In addition, the provider has an its own disaster recovery site in Rome to ensure RTO and RPO needed by Roma Capitale. Roma Capitale's main Data Centre is located in Rome.

The site in Rome is connected with the main site Milan with two dedicated MPLS (multi protocol label switching) VPN links (1 Gbps each link): this network connection is geographically differentiated (each link has its own bi-directional path from Rome to Milan) to enhance reliability, service availability and network resilience.

The Roma Capitale portal (www.comune.roma.it) is on a server located in Milan; while other services (including SUET service) are located in Rome. Front end user access (for common citizen) is to the Milan site, while employees gain access directly to the Rome infrastructure.

The mail server and related protection services are located in the Milan site. Security is enforced and monitored in by following elements:

• DDOS Mitigation services to protect Rome Capitale portal email services;
• Web application firewall to protect Rome portal;
• Network load balancer;
• Next generation firewall, both in Milan to protect Roma Capitale portal and email services (network perimeter security) and in Rome to protect each and every web service published by Rome Capitale portal homepage (Data Centre security) Many firewall layers exist.
• IPS, both in Milan and Rome, basically deployed following next generation firewall architecture.

The above top-level RP's were the starting point for the dependency analysis in both of the pilot cities. Each of the systems represented in each of these RP's were then the subject of further examination and analysis. Each were themselves the subject of a further Rich Picture, with each new RP drilling down through the functionality and connectivity of each application, system and network until all of those participating in the workshops were content that they had accurately captured, described and understood the functionality, connectivity and dependency of all of the applications, systems and networks. In a manner analogous to creating "Root Definitions", the workshop participants then wrote descriptions of their top-level RP's. CATWOE was then used to check the completeness of the RP's.

The key aspect of the CATWOE in terms of the SDA is that of the "Transformation(s)" identified in the RP. The transformation(s) are a description of the systems and networks that have been captured in the RP. These descriptions are the key to unlocking an understanding of how an organisation's networks and systems work and helps to enable the identification of key critical systems and the critical business processes that are conducted on those systems.

Key critical systems are either:

1. Those systems that are vital to the financial well-being of the organisation.
2. Those systems which collect, process or transmit sensitive or personal data or information, or via which systems sensitive or personal data or information can be accessed.

In addition to being used to check the completeness of an RP, CATWOE was also used to help identify key features, attributes, capabilities and vulnerabilities of a system, as is done in the context of the CS-AWARE SDA analysis.

Once this process was completed and in the following rounds of workshops, machine-drawn representations of the applications, systems and networks were created and further refined and checked for completeness by the workshop participants. For example, Fig. 2.9 shows an illustration of a dependency graph created by this process. Figure 2.10 highlights the assets of the graph that are involved in the information flows that a specific critical business process of this example organisation generates in day-to-day operation.

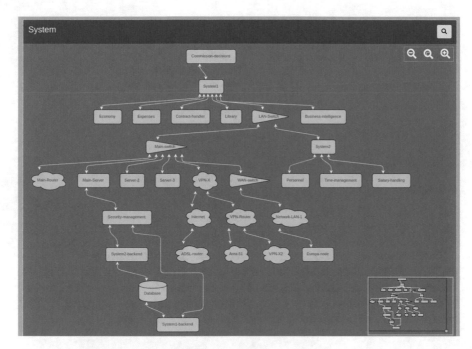

Fig. 2.9 CS-AWARE system and dependency graph

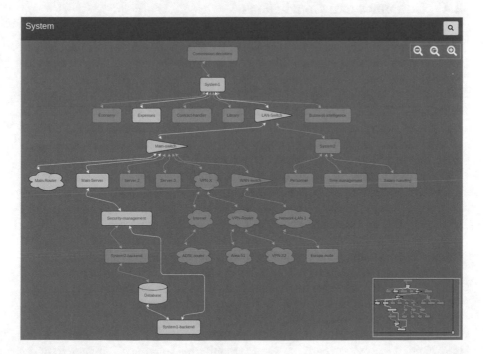

Fig. 2.10 Example of a business process information flow

Table 2.4 Illustration of a Monitoring pattern

Pattern name	Pattern parameters and ranges
Suspicious database modification attempt	Abnormal login time periods: Normal range (5 am, 9 pm) Max range (0 am–12 pm) User type Type: (admin, user) Number of users normal range (0;15,000) Max range (0; 50,000) Login session duration time: Logoff time—logon time Direction of search higher better If login rejections number >5 /h—suspicious Boolean: True/false

Furthermore, the table below gives an example of an illustrative monitoring pattern to be utilised for real-time monitoring of abnormal behaviour of the database utilised by a critical business process. The relevant monitoring parameters as well as the ranges are developed by the workshop participants from available log files based on the actual situation relevant for the use case (Table 2.4).

2.4   Outcomes and Conclusions

Our experiences with adapting and applying the SSM approach for SDA analysis in the LPA cybersecurity context have been overwhelmingly positive. We have experienced that the value of this type of analysis, and the learning effect that the participants in the LPAs experience, is much greater than the project has originally anticipated. While all the participants in the workshops are experts in their domain and area of expertise, the value of collaboration within the organisation to tackle a specific problem domain like cybersecurity from the holistic perspective is something that is not part of the usual activities of LPAs. The value of cooperation becomes more apparent the greater the size and complexity of a municipality is, since the organisational structures become more departmentalized and specialized. As a side effect of the analysis that was not anticipated to this level, municipalities who apply this approach can expect to develop a much more detailed and deeper understanding of their systems, procedures, their cybersecurity vulnerabilities and requirements implicit in their compliance with the GDPR. A more in-depth analysis of the value of SSM analysis will be given in Chap. 3, which represents an account of the CS-AWARE project from the LPA point of view, including considerations for structural and organisational changes to improve cybersecurity awareness based on the CS-AWARE approach and results.

In the context of being able to utilize the SDA results as a configuration basis for the technological cybersecurity awareness monitoring solution, the outcomes of the analysis have fulfilled the requirements in each respect. To reiterate, the main properties for the CS-AWARE solution to be able to LPA cybersecurity monitoring are:

- An asset and dependency graph: A structured representation of the asset and dependency analysis result, which can be used as an evolving knowledge repository of each asset's cybersecurity context.
- A model of information flows in the system and dependency graph: CS-AWARE enables the information flows for each identified critical business process to be modelled within system and dependency graph.
- Identification of observable parameters and the boundaries between normal and abnormal behaviour: The definition of CS-AWARE cybersecurity monitoring patterns requires the definition of observable parameters, as well as the boundaries between normal and abnormal behaviour observed at the identified monitoring points (log files).

We were able to gather and model this information in a structured format for both CS-AWARE pilot use cases. We were able to show that the significant difference of size and complexity of each use case (a mid-sized and a metropolitan area municipality) had no impact on the ability to achieve comparable results in both cases. Chapter 4, which details the different aspects of the technical part of the CS-AWARE solution, will show in more detail how this information is utilised to provide cybersecurity awareness.

References

Checkland, P. B. (1981). *1998*. John Wiley & Sons Ltd.
HUSAT Research Centre, Loughborough University. (1988). *Human factors guidelines for the design of computer-based systems*. Ministry of Defence Procurement Executive/ Dept of Trade & Industry.
Mumford, E. (1983). Participative systems design: Practice and theory. *Journal of Occupational Behaviour, 4*(1), 47–57.
Mumford, E. (1993). The participation of users in systems design: An account of the origin, evolution, and use of the ETHICS method. In D. Schuler & A. Namioka (Eds.), *Participatory design*. CRC Press.
Mumford, E. (2003). *Redesigning Human Systems*. IGI Global.
Mumford, E., & Henshall, D. (1983). *Designing participatively: A participative approach to computer systems design: A case study of the introduction of a new computer system*. Manchester Business School.
Trist, E. L., & Bamforth, K. W. (1951). Some social and psychological consequences of the long-wall method of coal-getting: An examination of the psychological situation and Defences of a work Group in Relation to the social structure and technological content of the work system. *Human Relations, 4*(1), 3–38. https://doi.org/10.1177/001872675100400101
Wood-Harper, A. T., Antill, L., & Avison, D. E. (1985). *Information systems definition: The multiview approach*. Blackwell Scientific Publications, Ltd..

Chapter 3
Story Telling

Jerry Andriessen and Mirjam Pardijs

3.1 Introduction

CS-AWARE stands for cybersecurity awareness. The solution that we want to provide to a municipality will support this organization to develop more awareness about their cybersecurity. In this chapter, we present our approach to understanding local CS-awareness at the beginning of the project, that is, before the system was designed and implemented. Our understanding of this user cybersecurity awareness would help the design of the CS-AWARE solution, from a sociotechnical point of view.

In October and November 2018, the CS-Aware project organised two one-day workshops at the premises of the municipalities that participated in the project. Participants in these workshops were department managers, system administrators and (internal and external) users of the municipal computer system network, including some of the technology services offered by the municipality. These workshops followed the second series of SDA workshops (see Chap. 2), during which the system administrators specified the complexities of using the municipality system network. The current chapter presents the rationale, the design and the outcomes of the one-day workshop, that we call *the story-telling workshop*.

The purpose of the story-telling workshop was to better understand participant experiences, in order to interpret their needs, roles and views about dealing with cybersecurity in their professional contexts. We decided that the best way to capture experiences was in the form of a story. By asking the participants to produce stories about their experiences with cybersecurity, based on (Kurtz & Snowden, 2003), we hoped to collect user perspectives in their own words and, at the same time, get an idea about user awareness of cybersecurity.

J. Andriessen (✉) · M. Pardijs
Wise & Munro, The Hague, The Netherlands
e-mail: jerry@wisemunro.eu

© The Author(s), under exclusive license to Springer Nature Switzerland AG 2022   45
J. Andriessen et al. (eds.), *Cybersecurity Awareness*, Advances in Information
Security 88, https://doi.org/10.1007/978-3-031-04227-0_3

In this chapter, we will first present the theoretical background on story-telling, and our approach on the analysis and interpretation of stories. For interpretation of stories, we propose a framework inspired by cultural-historical activity theory. This approach allowed us to dig deeper into the norms, rules, objectives and organizational constraints underlying participants' ideas and awareness about cybersecurity in their context. After that, we present the results of our interpretations for system administrators and system network users, separately, for the two municipalities. The final section discusses implications of the outcomes for the CS-AWARE solution, and it discusses the relevance of our approach for understanding user awareness in the context of designing complex systems.

3.2   About Collaborative Story Telling

Stories have characteristic elements (Propp 1927). They mention or presuppose a setting and include agents or actors. Transformations that the agent wishes to achieve in the circumstances of the setting are called goals. Stories have a plot, they include sequences of actions and events: things that actors do, things that happen to them, changes in the circumstances of the setting, and so forth. Particular actions and events can facilitate, obstruct, or be irrelevant to given goals.

Our experience of human affairs comes to take the form of narratives we use in telling about them. What people do in narratives is motivated by beliefs, desires, theories, values and other intentional states. People telling narratives tell about reasons and intentions, often as justifications, all of which can be interpreted and evaluated (Bruner, 1991). Story-telling as the creation of communicative structure and meaning has been widely used for a variety of purposes, including technology design and usability (Dyson & Genishi, 1994; Haste & Bermudez, 2017; Mar, 2018; Ojo & Heravi, 2017). Our approach is inspired by the idea of participatory narrative inquiry by Kurtz (2014), where the recounted events are seen to reveal the underlying experiences, emotions and perspectives of the story teller. In the stories we collected for CS-AWARE, we added the collaborative dimension: by asking small groups of people to work together on generating a story, we will be more likely to get their *joint* experiences, emotions and perspectives. These joint narratives about cybersecurity events and incidents in their work, could then be taken as reflecting the shared awareness of the participants, about cybersecurity in their organizational context, with respect to the ideas put forward in the narrative.

The picture below (Fig. 3.1, inspired by Kurtz, 2014, p. 44) shows that our approach to story-telling starts with an (individual) recollection (in terms of a problem, happening in some context, being resolved—or not—in some way) of some experience related to cybersecurity (upper triangle). We asked participants to prepare such a recollection before the workshop. These experiences were subsequently, during the workshop, elaborated in small groups, a process we call broadening and deepening (Baker et al., 2007). Topics in the story are broadened by adding more topics: we suggest to our participants to include the role of the organisation, the role

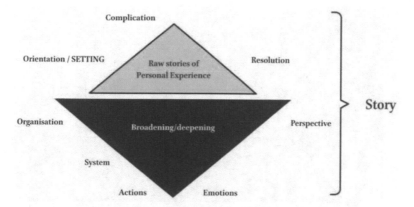

Fig. 3.1 Broadening and deepening of an experience gives the story

of the system, and the actions of the protagonist. Each topic can be deepened by going further on the path of perspectives, actions and feelings that characterise a topic (see Fig. 3.1). Broadening and deepening is done in small collaborative groups, because groups can help with reflection and elaboration, add multiple perspectives, and generally may enhance the range and depth of an experience into a story (Stahl, 2010). Our view of stories does not focus on creating complete narratives, but we aim for a set of coherent elements that characterise some event, within a context (orientation), the issue (the complication) and the outcome (the resolution).

We want to interpret the stories that are produced in terms of the organisational context in which the stories unfold. What we are interested in are the characteristic ways in which participants experience cybersecurity issues, and what this implies for their cybersecurity awareness. Our analysis therefore takes as a starting point the view of the user in the organisation, and the actions undertaken in that context, not the nature of the technological issue. Therefore, we need an interpretive framework that includes the organizational context in which the user is working, and how that constrains user experiences of technological issues, including their resolution. This framework is given by activity theory (Engeström, 1987, 2001), to which we will now turn our attention.

We consider this user with an issue as an activity system. An activity is seen as a system of human "doing" whereby a *subject* works on an *object* in order to obtain a desired outcome. In Fig. 3.2 below the object is depicted with an oval indicating that objects and their meanings evolve during activity. In order to achieve the objective, the subject employs *tools*, which may be external (e.g., an axe, a computer) or internal (e.g., a plan). As an illustration, an activity might be the operation of an automated call centre. Many subjects are involved in the operation and each subject may have one or more motives (e.g., improved supply management, career advancement or gaining control over a vital organisational power source). A simple example of an activity within a call centre might be a telephone operator (subject) who is modifying a customer's billing record (object) so that the billing data is correct (outcome)

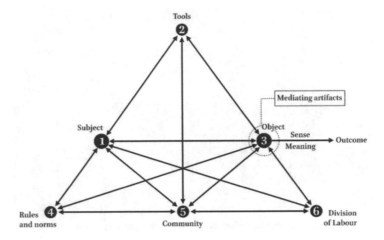

Fig. 3.2 The structure of a human activity system (after Engeström, 1987, p. 78)

using a graphical front end to a database (tool). An activity is a hierarchical structure, and its nature changes over time (Leont'ev, 1974).

For our purposes it is important to consider the tool, which 'mediates' between the activity and the object. "The tool is at the same time both enabling and limiting: it empowers the subject in the transformation process with the historically collected experience and skill 'crystallised' to it, but it also restricts the interaction to be from the perspective of that particular tool or instrument; other potential features of an object remain invisible to the subject..." (Foot, 2001).

An activity system further pictures an individual as part of a system characterised by particular *rules*, *communities* and *division of labour*. All activity is carried out within a social context, more specifically in a community of people, in our case the local public administration. The way in which the activity is situated in this context can be characterised by rules, which formally or informally mediate between subjects and a community, and by a division of labour, mediating between objects and the community.

Activity theorists argue that human experience is not a set of discrete disembodied cognitive acts (decision making, classification, remembering), and certainly it is not the brain; rather, consciousness is located in everyday practice: you are what you do. (Nardi, 1996). Nardi (1996, p. 5) argued that "activity theory proposes a strong notion of mediation—all human experience is shaped by the tools and sign systems we use." We may reflect on the meaning of awareness in a similar way: awareness is revealed in what people do, shaped in the context of their activity.

An activity system is usually not a smooth operation, it often characterises a current state of affairs in which tensions are operative, between the various components of the system. Tensions have a particular motivation in the activity systems approach, which is to identify contradictions in participants' work settings to help them change the nature of an activity to overcome those tensions (Engeström & Sannino, 2012; Langley et al., 2013; Marwan & Sweeney, 2019). For our purposes, we also focus on the notion of tensions, but here these tensions refer to underlying organisational inconsistencies and contradictions, that may frustrate people, but if people realise where their tensions come from, they may serve as a motor for change and improvement. Seen in that way, tensions may create awareness at the organisational level. Describing the tensions in an activity system may (partly) help us understand the main obstacles for cybersecurity in the organisation.

The story telling workshop involved small groups of participants from the municipality (system administrators and users from various departments) in collaborative efforts to produce meaningful stories about their cybersecurity experiences. These stories were presented and verified with all participants, including members of the CS-AWARE project. In the current report, the stories are reported and analysed in a CHAT- framework that aims to capture how organisations deal with cybersecurity issues, and uncover the main underlying tensions within these dealings.

3.3   Workshop 1

Present at Workshop I were 4 system administrators who had also participated during the three previous days in the SDA workshop (see Chap. 2). The local team leader had, in addition, engaged 10 additional public administrators, from several departments of the organisation. Four researchers from the CS-Aware project also took part, as moderators. Initially, all participants were divided in four groups of 3–5 members each. As the meeting evolved, some people left and a few others arrived, and the original groups mingled into different compositions (Fig. 3.3).

Three weeks before the workshop we had asked the participants by email to send us short stories of such experiences. With this preparatory assignment, we hoped to focus the awareness process on an individual level in order to give participants space for their own ideas. We received eight stories that guided our expectations of topics for the stories.

Fig. 3.3 Pictures from the story workshop

3.3.1 Workshop I: Procedure

After the explanatory talk, all participants from the municipality were (again) asked to individually generate an experience that involved some issue with the cybersecurity of their organisation. It was free for them to write it down or not. After 10 mins participants were asked to join their assigned group. The task of the group was to develop one of the individual stories, by adding topics to the experience and to provide these topics with more depth. The groups could discuss in their native language, but they were asked to put the elements of the story in English on a large sheet for plenary presentation (see Fig. 3.4). This phase took about half an hour. The plenary presentation of the first four stories took about 90 mins, as the presentations invoked a lot of discussion.

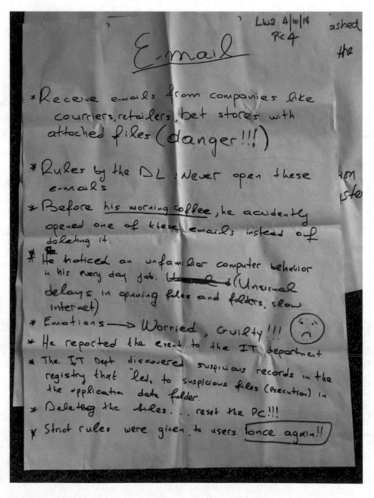

Fig. 3.4 Example of a generated story

Because the four stories of the first set were quite similar, albeit interesting in their own right, we asked system administrators to contribute different stories in the second round. This proved to be a difficult task, with a lot of discussion. This still resulted in another two stories, giving a total of six stories, which completed the session of 4 h.

The experience was not evaluated explicitly, but all participants may confirm a cooperative spirit with an open and inspiring discussion. Some tension arose during the second round of story-telling, because of a lack of clear ideas and criteria for new stories that were 'different' from the first round of stories. This was resolved by granting participants their time for brainstorming and collaborative negotiation.

3.3.2   Workshop I: The Stories

(For reasons of privacy, the gender and further information about the identity of the participants is not revealed)

Story 1: Mail for the Mayor   One of the tasks of this employee (E1) is to read all mail the mayor receives, which may comprise several hundreds of emails every day. In spite of strict regulations, E1 opened an infected email. This mail had no subject, which could have caused suspicion. After E1 opened the mail, files in each directory on his computer he accessed became inaccessible. The IT-Department was notified, they discovered a type of ransomware that had encrypted all the files in the directories E1 had accessed. The computer was restored in the IT-Department, where all files that were not yet accessed were retrieved, and the system was reinstalled, and the intact files were restored. Infected files could not be retrieved. E1 was initially terrified, feared the computer was lost, but now looks at it as a lesson to more strictly apply the rules for such cases.

Comment: The story is interesting, in the first place because it shows that regular activities that belong to someone's job description can already lead to dangerous situations. Secondly, the story shows that even though users may be very careful, generally aware of the possibility of danger, occasional mistakes may still slip through. The third interesting point is the role of the IT-Department: they resolve an issue that the user cannot possibly understand nor resolve, and as a result of which they remind the user about the strict safety rules. The emotional consequence is that users may be feeling insecure about using technology, and feeling guilty about having made a mistake.

Story 2: Inaccessible Network   This user (E2) works with various web-applications as regular part of his work. E2 suddenly was not able to find (on his computer) some other computers and printers on the network. The IT-Department found out this was produced by a worm-virus that caused crashing of a computer's browsing services. The infection travelled through the network to other users' pc's. The IT-Department found a system patch which was able to resolve the issue. It was installed on all computers on the network.

Comment: Again, we have a user for whom the work involves accessing various resources on the web that can be sources of compromise: social media, such as Facebook (also applications), chat-applications, google-drive, various web-pages. There clearly is a daily risk. The IT-department is the competent authority to resolve the issues. What happens if they are not available? Is there something else users should know, except following the rules? To be informed of recent issues? There may be a knowledge management issue, with respect to distribution and sharing of knowledge within the IT-department (although they discuss their case on a daily basis), but especially between that department and the other departments. Finally, patches are like pills and vaccines, but they may have side-effects.

Story 3: The Auto-CAD Virus An officer (E3) from the Urban planning department works with AutoCAD, a design tool. The tool was infected by a well-known virus that replicates in every folder with AutoCAD files. It can cause files becoming impossible to open and computer crashes, but often nothing happens, so, unlike the previous two cases, it is not always reported. Only after many restarts user E3 reported her problems: I cannot open a file, and my computer restarts. The IT-department easily detected the virus. They not only cleaned the infected machine (which means all the infected files, you have to go through all the folders, this already takes several hours), but also all other machines. But there is always the danger of users that took home a memory stick with infected files, and will bring this back later, even after a few years.

Comment: The virus takes advantage of full-access of some data storage repositories, and replicates in there, and then has possible access to other users of the network repository. External storage devices can also be a source of infection, especially USB-memory sticks that are not only used within the office. Sometimes AutoCAD files are exchanged with citizens. Using email for such exchange, or uploading on Google-drive is to be preferred. Users are recommended to scan the USB-stick. Following the rules is the only solution, but this requires all people to stick to the rules. Dangers are permanent and a virus may reappear after several years, when an old memory stick re-renters the system. Combating viruses requires the awareness of all people, not just a couple of them, and applies to all interactions with their computers, not only some of them.

Story 4: Opening an Email Of course, all users are careful users, in the departments. Many suspicious mails arrive, from companies, couriers, Nigerian Princes, etc. We know these can be dangerous. But before morning coffee, sometimes it happens that an employee (E4) opens an email and also its attachment. After a while E4 noticed delays in writing to files or folders and with navigation, suggesting a slow connection, and she realised she made a mistake, and felt very guilty about it. She reported to the IT-department nevertheless, and it was noted (informally) this was her second report. The IT-department discovered that there were suspicious records in the registry that led to suspicious execution files in the application data folder.

The pc was reset and the problem was solved. User E4 was again reminded of following the strict rules.

Comment: Another case of sabotage from an unknown source. Many people have stories like this. Every time the IT department was helpful, and every time the advice is to better follow the rules. Many users may feel guilty.

The next two stories are provided by people from the IT-department. We asked them to think about stories that were different, things you were not able to solve so quickly, or imaginary cases of what could happen if people are not careful enough?

Story 5: MS17-010 About 1.5 year ago, there was an attack on all Windows machines on the network. One user reported her pc was restarting immediately after initial start-up, and this happened again after 2 mins. The IT-officer tried booting with an external source, which seemed to work fine, so there was nothing wrong with the hardware. In the meantime, the same thing happened in another office. Within a few hours ten pc's were affected. Solutions were investigated by looking on social media (external to the municipality) for similar cases. The next day 40 users were affected. Many forums were explored on the network for solutions. What was found was that every infected computer used a (free) antivirus application that was not updated. Other pc's with a different antivirus system had no problems. Windows XP could not handle a specific procedure, (attack through a specific port) which caused the computer to crash and restart. The IT-officers looked for an existing Windows-XP security patch, but Microsoft had stopped supporting this version of Windows. Only much later a patch was released, because many other companies suffered from the same attack.

Comment: This is an interesting case of security awareness and sharing of information. The system department used social media (blogs and forums), for finding information shared by other security companies. This worked very well. Reports on Facebook of similar cases reveal the possibility that this form of sharing may be widespread.

Story 6: Hiring Temporary Support In our municipality we pretty much know each other, people have permanent working contracts that helped to survive the last 8 years of depression in Greece. However, during that same period there were also many new colleagues coming and going on temporary contracts (between 2 and 8 months). This increases the tension related to trust and openness of networked information and services. It is possible in many cases to retrieve database information in a readable excel format. It is hard to monitor and trust the behaviour of people we do not know very well, who work with some client computer.

Comment: In every networked system, people rely on each other and trust evolves over time about the reliability of people in maintaining the strict rules of cybersecurity. With more temporary contracts, it is more difficult to develop such trust, and the dangers for sabotage and theft increase. How to deal with such conditions? Give some people less permissions? It shows how much cybersecurity is a socio-technical and holistic problem.

3.3.3   Workshop I: Analysis

The six stories permit to depict two activity systems that each characterise activities with respect to cybersecurity issues from different points of view: that of the user of services (members of the departments of the municipality) and that of the IT-department in particular. We can easily understand that both activity systems interact in the case of cybersecurity issues. The views laid out in the stories seem coherent enough to warrant a valid interpretation. Our interpretations were checked with the users.

3.3.3.1   The Service-User Activity System

The service-user activity system that we describe is a generalisation, based on a limited number of stories that we have been collecting, but it should be said that the stories provided a coherent description. Like any activity system, it represents multiple voices (as in points of view, interests or traditions, positions), it has been shaped by the history of the unit and of the municipality, and can contain tensions and contradictions, which can be a motivation for change (Engeström, 2001, p. 137). Concerning our interpretation of such tensions, we have to be careful, they should be taken as possible sources of tension. Figure 3.4 depicts this user-activity system.

In this activity system, the *subject* is the generic user of techno-services has the *objective* of getting some (part of the) job done, by making use of a computer (the *tool*) as a regular part of the profession, but is faced with some obstacle so that the regular activity cannot be performed. This gives rise to a shifted objective: the resolution of the technical issue. Characteristically, this user does not have the knowledge (tools) to resolve the issue, so the user needs the IT-department for this. Enter activity system 2, depicted below in Fig. 3.5, and now we can say there are two systems operating with the same objective: resolution of the issue, but according to different principles, and with different tools.

In the case of this municipality, the subject is part of the *community*: all users within a department, or in our case, all users in the municipal organisation. These users probably interact, but we do not know the extent to which these interactions concern cybersecurity. The stories show that our users may have some generic awareness of the possibility of risks, and such risks may appear in some discussions between users, but there is probably no general (formal) practice for sharing cybersecurity issues with all members of the community, or even within a particular department. This means that some issues may reappear, with a different user. We find examples of the tensions related to this lack of transparency in all stories.

In principle users all share the *rules*: there is a set of safety-regulations set out by the IT-Department for secure cyberbehaviour. The most important rule is: "all users are warned not to open e-mails from unknown senders or without a subject or with attached files in .zip or .exe format". We suppose this list of regulations covers all situations presented in the first four stories, and the list probably is updated with

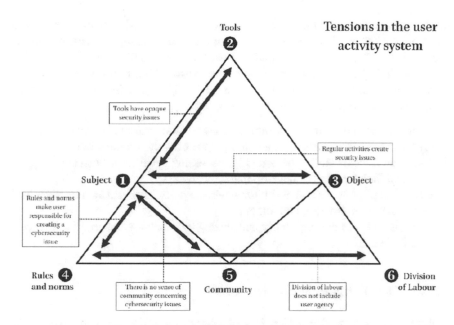

Fig. 3.5 The service user activity system characterises users of services from municipality 1

every new case, when necessary. A general rule is that all cybersecurity issues should be reported to the IT-Department, who have the knowledge to adequately deal with them. The rule is for the benefit of the community: if I do not report then the community is endangered. But this rule also creates possible tension for the user: if I report then people may think I am a sloppy user (whose behaviour is bad for the community). Finally, there appears to be no sharing rule that applies to the community: what is shared about certain events within the community?

The *division of labour* is very clear: all cyberissues will be handled by the IT-Department. They are seen as competent, reliable, quick, and are especially good at estimating the risk for the organisation and taking the necessary measures. On the other hand, our user knows to be seen as a risk, a possible source of mistakes, who needs to stick to the regulations to avoid making these mistakes. The procedure is clear: the user with the issue hands over all responsibility, and then will be returned a repaired machine as soon as possible. If needed, other machines within the system will be handled as well.

This leads to the final part of the activity system, which is the *tool*. The user, any user, probably understands how to operate the machine as an instrument for doing some jobs. As a tool, it functions in so far it is needed. Part of the meaning of the tool is doing the job correctly. If this fails, then the machine is not a tool anymore, but an obstacle. What is missing for the user, as part of the tool, is the knowledge about the system and the system network. We already said, above, the rules and the division of labour do not assume that users possess any agency for remediation. In the current system, they are not supposed to know anything about the possible relationship between their cybersecurity and their way of using tools. It is an open question what type of awareness might support better handling of issues from the user side, or less tensions when handing over the tool, and if this is desirable.

We see in Fig. 3.5 our interpretation of the tensions within this user activity system, under threat by a cybersecurity issue. The following tensions were identified between the various components of that system:

1. Users do not understand the security issues of their system and system use
2. Cybersecurity issues often are a consequence of daily routine work
3. Users are not expected to be active in discussing cybersecurity issues and policies, and therefore do not develop much awareness
4. Users feel responsible and guilty for a cybersecurity issue and are therefore hesitant to share information with others
5. There is no sense of community concerning cybersecurity issues, it is everyone for him/herself.

3.3.3.2   The IT-Department Activity System

In this activity system, the *subject* is a member of the IT-Department (five people) that handles all issues with cybersecurity in the municipality. The department has one head, respected for his precision and expertise, the other members are equal. They are member of (at least) two communities: the first is the municipality that they work for, and maybe some smaller municipalities in the area. This community is the main client of the IT-Department, and the wellbeing of this community is the main motivation for their services. There is frequent exchange of issues, ideas and solutions between the five employees, but (as we assume) not in a systematic manner with the other departments. The second community is that of IT-specialists with a similar task description, including cybersecurity response teams, task forces and competent authorities. This second community is important for development of expertise and for consulting possible solutions for issues, but also for conveying information about (new) threats to cybersecurity. Sharing activities with the community of professionals was not explicitly addressed in the story-telling, albeit there was clear mention of consultation of blogs and forums where active communication within this community mostly takes place. It is unknown how frequent the members of this Department exchange feedback with other members of the community of their profession.

The *objective* that is shared with the user activity system is the resolution of cybersecurity issues, meaning the wellbeing of all users. Perhaps in a more general sense such objectives give rise to development of *tools*: experiential knowledge for knowing how to act, including where to find knowledge, expertise about networks and systems, safety rules and antivirus solutions such as patches and cleaning software. Tools are main the domain of expertise for this activity system. However, it appeared from the stories that the municipality uses outdated versions of MS-Windows.

The *rules* (and norms) the IT-Department maintains are monitoring safety behaviour, and providing high quality services to the users. It seems they are doing fine, but they are alerted by the users themselves, when a user reports a problem, and do not (often) prevent issues before they actually occur, through monitoring network activity. Moreover, the IT-department is expected to resolve all issues, and users have a tendency to held the sysadmin responsible for malfunctions of their technology.

Concerning the *division of labour* (already discussed in the section about users) they raised an additional point in story 6: due to increasing flexibility on the job market and availability of jobs, it may be possible that more and more people will be employed on a temporary basis. These people may have different norms, with respect to safety regulations, but also with respect to the importance of the community.

We see in Fig. 3.6 our interpretation of the system department activity system, acting in case of a cybersecurity issue. The picture focuses on the tensions that we identified between the various components of that system:

1. System administrators are held responsible for an employee's issues, they learn from their work, but employees do not
2. There is no sharing policy for cybersecurity issues within the organization, only within the IT-department
3. The organization works with outdated versions of MS-Windows, and therefore frequently has to resolve outdated issues or issues that cannot be resolved with this system

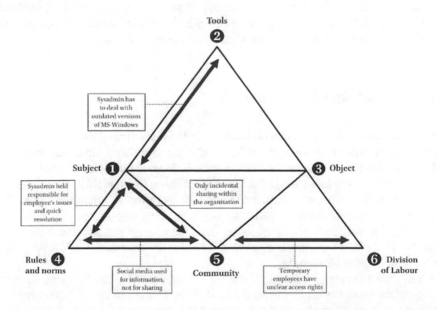

Fig. 3.6 Tensions in the IT-Department activity system, for municipality 1

4. There is an issue with access rights for temporary staff
5. The system administration uses information from reliable social media sources, but has no policy for sharing with this community

3.4 Workshop 2

The municipality for workshop 2 is a very large capital. Present at workshop 2 were a number (between 10 and 15) of system administrators who had also participated during the three previous days in the SDA workshop (see Chap. 2). The role of these participants within the municipality was not only that of system administration, they could also be managers of contracts, or representatives of private parties that were responsible for a particular component of the system. The local team leader had engaged three additional public administrators, from the department that handled European contracts. Incidentally, some other users participated, but there was frequent coming and going of people. Four researchers from the CS-Aware project also took part, as moderators, and we gratefully acknowledge the contributions from the local team leader of workshop 1. Initially, all participants were divided in four groups of 3–5 members each. As the meeting evolved, some people left, then came back or not, and the original groups mingled into different compositions. The procedure was similar to the one reported in the section about workshop 1.

About 4 months before the workshop, we asked for participants by email to send us short stories of cybersecurity experiences. With this preparatory assignment, we hoped to focus the awareness process on an individual level in order to give participants space for their own ideas. We did not receive any stories.

3.4.1 Workshop 2: Procedure

After the explanatory talk, all participants from the municipality were asked to individually generate an experience that involved some issue with the cybersecurity of their organisation. It was free for them to write it down or not. After 10 mins participants were asked to join their group. The task of the group was to develop the story by adding topics to the experience and to provide these topics with more depth. The groups could discuss in their native language, but they were asked to put the elements of the story in English on a large sheet for plenary presentation. The story-construction phase took about 45 minutes. The plenary presentation of the first three stories took about 30 minutes.

The group appeared to be full of stories, so, after a coffee break, we asked them to generate more stories in a second round. This resulted in another three stories, giving a total of six stories, which completed the session of about 3 h.

The experience was not evaluated explicitly, but all participants may confirm a cooperative spirit with an open and inspiring discussion.

3.4.2 Workshop 2: The Stories

(for reasons of privacy, the gender and further information about the identity of the participants is not revealed)

Story 1: Phishing An email (from an external user) contained the image of the opening page of the municipal internet, asking for user credentials for verification purposes. A few dozen of users provided their credentials to a fake URL. After a few hours Internet Services (The IT-dept.) was aware of the danger and they acted to neutralise the danger (localise the infected pc's from system logs, block their accounts, reset passwords, blacklist the fake URL). It is unknown what happened with the stolen information, nor what was the perception of the users who were compromised. The attitude seems to be that this can happen to anyone. Also, there was no follow-up by Internet Services. So, it may well happen again. This especially true since it is easy to make up an email address if you know the name of an employee of the municipality.

Comment: This story was produced by a group of system administrators who imagined themselves as users. Therefore, it is hard to assume the full user viewpoint, and this analysis will be incomplete. This story is a classical phishing case, and some users were aware of the danger immediately. Others were not, but it is unclear what numbers they represent. It appears there was no follow-up, as when other users are immediately informed of this danger, or later when all users are informed about recent dangers. We do now know to what extent user regulation concerning suspicious emails are spread amongst the employees of the municipality. We might have to deal with serious knowledge management problems that will not be resolved with a better virus detection system, in addition it raises questions about what the awareness services of such a system need to accomplish.

Story 2: Fake News In the news media it was written that the personal identity of many citizens that registered on the portal was stolen. This included the Mayor herself! The Internet Services were alerted by a journalist about this news. After a lot of time spent on inspecting the system and the log files, it could be affirmed that the news was fake, and there had been no stealing of identities.

Comment: amongst the many dangers, fake news, or manipulated news is a more recent one. This is hard to handle by traditional virus scanners because it involves confirmation that there is nothing wrong with the system. It only works when the system is sure of itself: I detected nothing wrong! As we will learn later, the system in this municipality cannot be certain, because its updates are erratic. Another option would be the possibility to question the system, to ask the system questions about a general class of certain symptoms that have not previously been diagnosed as dangerous. We may wonder what the interest of such fake news attempts is, probably disrupting the image of the town municipality as reliable.

Story 3: Forbidden Websites   This user group generated a story about the local restrictions to visit certain websites. This particular user works at a department responsible for managing European contracts. The problem that is frustrating them, and probably many others, is that some websites from the EU are forbidden by system administration. The procedure for employees to get access is for the head of the department to write an official request for access to certain websites by certain people. This helps, but only for the people who have been listed in the request. All of this is interpreted as erratic by our user: is there any real policy behind this?

Comment: The issue seems to be that the policy behind allowing or restricting websites within the municipality is unclear, and probably erratic, in the sense that it may be the result of multiple policies operating at the same time, including individual decisions. As a consequence, some things are allowed and some things are not allowed. For the user, this is highly unsatisfactory and confirms a lack of trust, in addition to searching for individual solutions rather than for those that relate policy issues in general. There clearly is a transparency problem for the IT-Department AND for individual users.

Story 4: Password Reset   This is the classical story about somebody who called the IT-Department to reset his password, because he forgot what it was. The IT-employee asked for the user's personal name (but not for the personal codes only known to this user), which was Mario Rossi. The password for Mario Rossi was then reset. Unfortunately, it turned out that there were 3 Mario Rossi's working for the municipality (it is a very common name and there are more than 20.000 employees), and the wrong Mario Rossi's password was reset. This only appeared when Mario Rossi 1 called again telling he could still not enter his email. This was now settled, but then there still was the issue of Mario Rossi 2, whose password was also reset, by mistake. This appeared to be an employee who had retired 2 months ago. This revealed another problem: when people leave, the IT-department is not informed.

Comment: This appears to be another problem for the IT-department, that their own employees do not always follow procedures. In addition, the question is how to solve the problem what happens when someone leaves. Again, we do not know what the user experienced, as this story was produced by the system administrator group.

Story 5: Web Conferencing   A large municipality can save a lot of time and money when its employees do not have to travel to meet their colleagues at other places, or even abroad, but instead can use web conferencing, such as Zoom or WebEx. This is formally not allowed by this municipality. Informally however, employees can use their Phone on the G4 network, or they can work from home. This is what most people do. There is a problem when files on personal laptops are compromised and transferred, or when files from compromised USB-sticks are transferred through the conferencing app.

Comment: This is a similar story as story 3, and it adds to the lack of trust in IT-policy making. There was no sign of understanding by this user why web conferencing was forbidden by the municipality. On the one hand, many users from other public organisations may recognise this issue. More transparency of policy rules and more interest in user perspectives may well contribute to less issues and greater trust.

Story 6: Patching This story confirms there is no policy for patching and updating from the side of the IT-department, and this is not seriously considered by management. It is not verified if users use the latest versions of software, and if all updates and patches are installed. There is no test environment operative. The management says there is a lack of money.

Comment: At the management level there seems to be a clear lack of attention to security updates and patches. Will awareness in this respect help anything?

3.4.3   Workshop 2: Analysis

Our analysis takes as a starting point the view of the user in his organisation, not the nature of the technological issue. Therefore, we need a framework that does justice to the organisation in which the user is working, and how that relates to how a user experiences the technological issues, including their resolution. We consider this user with an issue as an activity system. This is further explained in Sect. 3.3.2. It should be noted that for the interpretation of the stories in workshop 2, there is a lack of stories from actual users. Therefore, our interpretation of the user activity system is incomplete and tentative.

3.4.3.1   The Service-User Activity System

The service-user activity system that we describe is a generalisation, and might be different for each department in the municipality. These differences have not appeared, given the limited number of stories that we have been collecting. Like any activity system, it represents multiple voices (as in points of view, interests or traditions, positions), it has been shaped by the history of the unit and of the municipality, and can contain tensions and contradictions, which can be a motivation for change (Engeström, 2001, p.137). Concerning these tensions, we have to be careful, they should be taken as possible sources of tension.

The *subject* of the activity system is the generic user of techno-services with the *objective* of doing some job, but is faced with some obstacle so that the required activity cannot be performed. This is the initial issue and this gives rise to a shifted objective: the resolution of the technical issue. Characteristically, this user does not have the knowledge (tools) to resolve the issue, the user needs the IT-department for this. If we compare the stories generated by participants from municipality 2 with those from municipality 1, we see that the issue is different. In the current municipality, the users seem more frustrated by institutional obstacles in their professional activity, in addition to technical ones, because of the existence of seemingly ad-hoc regulations that, for example, prevent their access to websites or the use of applications needed as a regular part of their job. Characteristically, users try to get around this issue, either by negotiating with the IT-department, or by using their own technology. The first solution does not resolve the general issue for the municipality, but only for assertive users. The second solution may give rise to additional cybersecurity issues.

We know that users also encounter the more regular cybersecurity issues, such as phishing or viruses like in the stories from municipality 1. An informal questionnaire issued by the municipality team-leader, with about 44 respondents shows the distribution of such issues. These were not raised during the workshop, we do not know to what extent users regard these as very problematic. We also do not have much information about the professional communities in which users are functioning.

The core of frustrations by the users reveals some imperfections of the *rules* and norms. Awareness of safety rules is beneficial for users, and, as was the case in municipality 1, users acquire such awareness by trial and error, and through interactions with others. In the current case, regulations are experienced as lacking any logic, or this logic is not understood. There is a clear tension between the professional needs of regular users, and the rules and regulations that are supposed to support their sense of security. This relates to the norm that rules are to be bypassed. This lack of trust in safety regulations may also relate to lack of trust in the policies of the organisation.

The *division of labour* amounts to users sorting out their technical issues, if possible, with the support of the IT-Department. The IT-department are seen as part of the organisation, hence there maybe is not much doubt about their competencies, but there is doubt in their procedures and the application of policies. Security issues seem less important for users than the frustration of not being able to do your job. As a consequence, we think that their awareness of cybersecurity issues on a daily basis is quite limited.

The user, any user, usually understands how to operate the machine as an instrument for doing daily jobs. As a tool, it functions in so far it is needed. Part of the meaning of the *tool* is doing the job correctly. If this fails, then the machine is not a tool anymore, but an obstacle. What is missing for the user, as part of the tool, is the knowledge about the system regulations. It is an open question what awareness might support better handling of issues from the user side, or less tensions when meeting a technical issue, and if this is desirable.

We see in Fig. 3.7 our interpretation of the tensions within this user activity system for municipality 2, under threat by a cybersecurity issue. The following tensions were identified between the various components of that system:

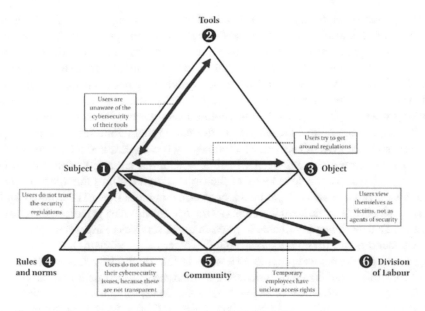

Fig. 3.7 Tensions in the user activity system, for municipality 2

1. Users do not understand or trust the cybersecurity regulations of their organisation
2. Users look for ways to circumvent the constraints that they encounter,
3. Users may create additional security issues
4. Users experience a lack of transparency of cybersecurity issues, so their sharing is limited, and the community (of other users in the municipality) remains uninformed about issues and their potential dangers
5. Users see themselves as victims of failing policies.

3.4.3.2 The IT-Department Activity System

Here, the *subject* is a member of the IT-Department (size unknown) that handles all issues with cybersecurity. The department in a larger sense is a very complex structure within a very large municipality. Many services have been privatised, and as a consequence, no-one has full overview or full responsibility over the network. It is unclear how communication between the various stakeholders takes place, or what communities in a smaller or larger sense they are involved in. We have learned from another workshop (see Chap. 6) that the department manager is the spider in the web, overseeing and directing most activities, and, especially, communicating between the various service providers and IT-specialists.

The *objective* that is shared with the user activity system is the resolution of cybersecurity issues. Perhaps in a more general sense such objectives give rise to development of *tools*: experiential knowledge for knowing how to act, including where to find knowledge, expertise about networks and systems, safety rules and antivirus solutions such as patches and cleaning software. Tools are main the domain of expertise for this activity system. Unfortunately, there seem to be insufficient funds for maintaining recent updates in the system network (This may be different for various private subcomponents in the network). Dangers for users are always lurking when they have to work with outdated versions of system software.

The *rules* (and norms) are adapted to the notion that within the huge municipal organisation the system network as a whole is too complex, and we should accept imperfection to the extent that not all policy regulations are in the interest of cyber-security. Obviously, the complexity of the organisation requires that many services are outsourced to smaller departments or external organisations. There may also be some lack of trust in the good intentions of the organisation management, but it looks like the IT-department is still able to provide quick services to users with problems.

Concerning the *division of labour*, the IT-Department maintains are for monitoring safety behaviour, and providing services to the users. The organisation of support for users is facilitated by a ticketing system, because resolution of almost any issue requires involvement of many specialists. Many support activities for users with issues are provided on an individual basis, and often there is no follow-up (if the user does not call again, it is assumed the issue is resolved), and there may be no spread of information to other users in case of a security issue. However, there may be employees within the department who engage in some form of information sharing with other employees.

An additional point was raised in story 4: who is responsible for the administration (access or removal) of users? Isn't it a problem when users leave the organisation but may still have full access to the system after several months? This may relate to the issue of knowledge management, to users, who seem to have less concern and a lack of awareness for cybersecurity issues, and to their own department, in which the culture favours general regulations (too many) but individual solutions (not shared).

The conclusion that cybersecurity is not the first concern of the IT-department is reinforced by the following observations: (a) there is no concern for the wellbeing of the users: were they helped, what happened with their issue, what happened with the data they lost?; (b) there is no follow-up after an issue has been resolved, not to all users, not (it seems) to others within the department; an issue can easily happen again; (c) repair and support is on an individual basis, it is unclear if there are policies, and mistakes can be made without clear personal consequences; (d) there are no clear procedures in place for users that have left the organisation; there seems to be no management of users that do not work for the municipality having access to sensitive information.

The second activity system (Fig. 3.8) therefore reveals quite general sources of tension, that do not relate to the user as an individual (who is now represented as the community of users), but more to organisational matters. Potential tensions between components in this activity system are reciprocal:

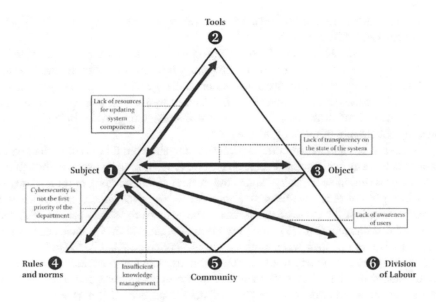

Fig. 3.8 Tensions in the IT-Department activity system, for municipality 2

1. Lack of resources so system components are not always updated, and therefore unsafe
2. Lack of transparency on state of the system and the organisation, so stakeholders are not aware of the current status of the system
3. Cybersecurity is not the first priority of the system department
4. Lack of awareness of user behaviour or user satisfaction, so users can arrange their own solutions which are not always safe
5. Lack of knowledge management, so there is no systematic approach to sharing and availability of cybersecurity relevant information

3.5 Main Conclusion

We identified some sources of tensions in the user- and system administrator- activity systems, in the stories collected during the storytelling workshops in two municipalities. Although it may seem there are some similarities between the municipalities in this respect, in fact we are talking about very different systems for dealing with cybersecurity.

When we use the term "users" it should be clear we refer to users of services, that is the professionals who work for the municipalities in any of the departments. When we speak about IT-administrators, or system administrators, we refer to the professionals who work in the IT-department of the municipalities. Some

IT-administrators are directly involved in resolving cybersecurity technical issues, others have administrative roles.

From what we have seen in the story telling workshop, the internal risks of uninformed, frustrated, former, current, or temporary new users are significant. Internal threats to local cybersecurity are a very clear challenge for these public administrations. This challenge is in most cases not caused by malicious intentions, but by significant lack of awareness, leading to misunderstanding, risky behaviour, and insufficient sharing of relevant information.

A more general conclusion that is confirmed again from this small collection of stories is that any technological solution requires considering its users. Although the CS-Aware solution specifically targets the system administrators, and their awareness in a general sense, we think that in order to have any impact, this awareness needs to consider the users of the services, the policies for safety regulations, and the potential dangers that are part of any internal organisation. We have seen examples of what can happen when policies are not coherent. Also, we highly recommend for system administrations to include the users in any feedback loop, to help them understand the reasons for their issues with cybersecurity, and to make them realise the consequences of some types of behaviour, even if it was someone else who experienced an issue.

References

Baker, M., Andriessen, J., Lund, K., Van Amelsvoort, M., & Quignard, M. (2007). Rainbow: A framework for analysing computer-mediated pedagogical debates. *International Journal of Computer-Supported Collaborative Learning, 2*(2–3), 315–357. https://doi.org/10.1007/s11412-007-9022-4

Bruner, J. (1991). The narrative construction of reality. *Critical Inquiry, 18*(1), 1–21.

Dyson, A. H., & Genishi, C. (Eds.). (1994). *The need for story: Cultural diversity in classroom and community*. National Council of Teachers of English.

Engeström, Y. (1987). *Learning-by-expanding*. Cambridge University Press.

Engeström, Y. (2001). Expansive learning at work: Toward an activity theoretical reconceptualization. *Journal of Education and Work, 14*(1), 133–156. https://doi.org/10.1080/13639080020028747

Engeström, Y., & Sannino, A. (2012). Whatever happened to process theories of learning? *Learning, Culture and Social Interaction, 1*(1), 45–56. https://doi.org/10.1016/j.lcsi.2012.03.002

Foot, K. A. (2001). Cultural-historical activity theory as practical theory: Illuminating the development of a conflict monitoring network. *Communication Theory, 11*(1), 56–83.

Haste, H., & Bermudez, A. (2017). The power of story: Historical narratives and the construction of civic identity. In *Palgrave handbook of research in historical culture and education* (pp. 427–447). Palgrave Macmillan. https://doi.org/10.1057/978-1-137-52908-4_23

Kurtz, C. (2014). *Working with stories in your community or Organization* (3rd ed.). Kurtz-Fernhout Publishing.

Kurtz, C. F., & Snowden, D. J. (2003). The new dynamics of strategy: Sense-making in a complex and complicated world. *IBM Systems Journal, 42*(3), 22.

Langley, A., Smallman, C., Tsoukas, H., & Van de Ven, A. H. (2013). Process studies of change in organization and management: Unveiling temporality, activity, and flow. *Academy of Management Journal, 56*(1), 1–13. https://doi.org/10.5465/amj.2013.4001

Leont'ev, A. N. (1974). The problem of activity in psychology. *Soviet Psychology, 13*(2), 4–33. https://doi.org/10.2753/RPO1061-040513024

Mar, R. A. (2018). Evaluating whether stories can promote social cognition: Introducing the social processes and content entrained by narrative (SPaCEN) framework. *Discourse Processes, 55*(5–6), 454–479. https://doi.org/10.1080/0163853X.2018.1448209

Marwan, A., & Sweeney, T. (2019). Using activity theory to analyse contradictions in English teachers' technology integration. *The Asia-Pacific Education Researcher, 28*(2), 115–125. https://doi.org/10.1007/s40299-018-0418-x

Nardi, B. A. (1996). Activity theory and human-computer interaction. In B. A. Nardi (Ed.), *Context and consciousness: Activity theory and human-computer interaction* (pp. 7–16). MIT Press.

Ojo, A., & Heravi, B. (2017). Patterns in award winning data storytelling: Story types, Enabling Tools and Competences. *Digital Journalism*, 1–26. https://doi.org/10.1080/21670811.2017.1403291.

Propp, V. (1927). *Morphology of the Folktale*. Trans., Laurence Scott. 2nd ed. Austin: University of Texas Press, 1968.

Stahl, G. (2010). Guiding group cognition in CSCL. *International Journal of Computer-Supported Collaborative Learning, 5*(3), 255–258. https://doi.org/10.1007/s11412-010-9091-7

Chapter 4
The Design of CS-AWARE Technology

Alexandros Papanikolaou and Kim Gammelgaard

4.1 Introduction

This Chapter focuses on the technological aspects regarding the implementation of the CS-AWARE solution, which provides system administrators with cybersecurity awareness about the information system they are in charge of, by analysing the information found in the log files of their most critical systems and visualising the results in an appropriate manner. In this way, system administrators are quickly informed whether there are indications of suspicious activity occurring in their systems and they also receive recommendations or suggested actions to take for specific instances of the aforementioned issues. As a result, system administrators do not need to spend time analysing and correlating potential indicators to form an opinion on the cybersecurity status of their systems. They can use this time for further understanding the information and the potential decisions presented to them by the system. Furthermore, by collecting and analysing publicly-available cyberthreat intelligence, the CS-AWARE system is able to deduce whether there are cyberthreats in the wild that could harm a specific information system it monitors and issue the necessary warnings accordingly.

The functionality that CS-AWARE provides aims to help organisations deal efficiently with the cybersecurity challenges their information system is faced with, taking into consideration that they usually lack both the expertise and resources to properly address these issues. The chapter presents the deployment of the

A. Papanikolaou (✉)
InnoSec, Thessaloniki, Greece
e-mail: a.papanikolaou@innosec.gr

K. Gammelgaard
RheaSoft, Aarhus, Denmark
e-mail: Kim@rheasoft.dk

CS-AWARE system at two pilot Local Public Administrations (LPAs) of different sizes, as examples of organisations that lack resources to allocate for dealing with their cybersecurity issues. Nevertheless, the solution could easily be applied to SMEs as they share similar characteristics with LPAs as far as their information system and resources (both human and financial) are concerned.

The individual seven components of the CS-AWARE solution are initially presented, explaining their functionality and role in the overall system. Then, the CS-AWARE Framework is presented, exhibiting how the aforementioned building blocks interact among them to achieve the desired system functionality that essentially involves analysing data, deriving cybersecurity awareness and conveying it to the user.

The implementation phase presents the various challenges related to the complexity and heterogeneity of the systems that had to be addressed from a technical perspective, as well as the main key points where decisions had to be made in order for a viable solution to be produced, given that there were developer teams from different companies/institutes involved, with different backgrounds, expertise and work culture. The integration and deployment phase was equally challenging, since all the aforementioned building blocks had to communicate among them and also with the designated information system nodes of each one of the two pilot LPAs (Municipality of Larissa, Greece and Municipality of Rome, Italy).

Finally, the user interface design is presented, demonstrating how it evolved from its initial conception to its final "look and feel", as well as the functionality and the information it presented to the user. This was the result of several rounds of iterations, where feedback was collected from users, was analysed and incorporated into the user interface through appropriate modifications. Moreover, the implementation of certain changes in the user interface triggered modifications in the functionality and/or the information provided by the other building blocks, so as to offer the highest possible level of cybersecurity awareness to the user.

4.2   The CS-AWARE System Architecture

In this Section the architecture of the CS-AWARE system is presented, starting from the description of the main components it consists of. Then, the presentation of the CS-AWARE framework follows demonstrates the logical grouping of the said components, as well as the information flows among them during their operation.

4.2.1   Description of the Main Components

Before going into the details of how the CS-AWARE was implemented and the related decisions that had to be made during this course, it is worth presenting in brief the various components and their purpose in the integrated system. The seven modules comprising the CS-AWARE system (System and dependency analysis, Data collection and storage, Data analysis, Multi-lingual semantics support, Data visualisation, Cybersecurity information exchange, Self-healing) are presented in the next sections.

4.2.1.1 System and Dependency Analysis

By applying the Soft Systems Methodology (SSM, see Chap. 2), rich pictures were drawn relating to the organisation's systems and several problem situations were expressed in those pictures. The analysis of these rich pictures helped in identifying three main categories of systems:

- Important systems that contained personal or sensitive data and had therefore to be adequately protected.
- Systems that contained mission-critical data for the operation of the organisation (e.g. the accounting and financial system).
- Points in the network / systems architecture where it could be possible to collect data for analysis (e.g. log files) so as to identify anomalous events that would signify potential security breaches.

During the analysis phase, the following potential data sources were identified at the following levels:

- Database level. Most modern database systems have built-in monitoring and auditing functionality based on log files, which can be utilised for monitoring all database operations (insert, modify, delete).
- Application/service level. Auditing logs related to applications or services that can be utilised for observing the application/service behaviour.
- Network level. Most modern network equipment provides excellent auditing and monitoring capabilities that can be used as input to the analysis engine.
- Security appliances (software or hardware). Firewalls, Intrusion Detection / Prevention Systems (IDS/IPS), SIEM systems, anti-malware software can provide information about security incidents than can be used to obtain a clearer, security-oriented view of the systems' state.

Once the aforementioned analysis has been completed, the obtained information is input to the Dependency Mapper tool, to make it available in a machine-readable form. Every identified node is entered as an asset that also contains all necessary information that can be used for uniquely identifying it as a node (e.g. hostname, IP address), as well as with respect to the type and the version of the operating system running on it. Furthermore, any interconnections to other assets of the system are also added to this information. As soon as this process has been performed for every asset that was identified as a critical node, a system graph can then be produced that shows all the critical nodes and their interconnections. This mapping information is made available in structured form (JSON) to any other module that may require it for fulfilling its operation.

4.2.1.2   Data Collection and Storage

This module involves a data collection framework, able to aggregate and ingest data from multiple sources in diverse formats (e.g. TXT, CSV, JSON and XSL) and in diverse operating systems (Windows, MacOS and Linux). Moreover, the framework has been designed to be flexible and allows the integration of additional sources, should it be required. The collected data is fed into the CS-AWARE thread detection engine for analysis to detect threats or provide pointers for mitigations. Four main sources were used for gathering data:

1. Logs from servers, databases, applications and network devices from within a Local Public Administration's (LPA) systems.
2. Information about packages and software installed on servers.
3. Cyber threat intelligence from specialised websites and feeds.
4. More general cybersecurity-related notifications and warnings collected from social networks.

It is worth pointing out that the first two sources may contain personal and sensitive data, which therefore demands for special handling, so as to be compliant with applicable laws and regulations, such as the GDPR. The latter two sources contain publicly available data. Regarding the requirement for GDPR compliance, suitable pseudonymisation or anonymisation is performed at the data source before it is sent to the cloud data collection sink. All data transmissions take place via secure SSH/SFTP channels between the source and the data sink, so as to preserve the data's confidentiality (particularly in the case of sensitive data), integrity and authenticity.

Cyberthreat intelligence (CTI) contains a lot of information that is normally used to understand a threat actor's motives, targets and attack behaviours. This information is made publicly available in a structured form, using appropriate protocols, such as STIX (the XML-based version 1.x and the JSON-based version 2.x). For the purposes of CS-AWARE, specific STIX fields are selected from which data is extracted and provided as input to the threat detection engine.

Social networks are also exploited as, under certain circumstances, they contain cybersecurity-related notifications and warnings. For the purposes of CS-AWARE, certain user accounts related to cybersecurity are followed and data is collected from their posts. Hence, in case a cyberattack is in progress, it is quite probable that some relevant information will be published via the aforementioned accounts, which could help in shielding against or recovering from this attack.

4.2.1.3   Data Analysis

This is the threat detection engine of the CS-AWARE system, where data is analysed, correlated and matched against patterns that would identify specific threats and/or suspicious activity. All kinds of data are exploited and correlated, so as to maximise threat detection, such as:

- Vulnerability information is matched against assets. For instance, look for known vulnerabilities related to specific OS versions.
- Threat information is combined with logs and assets. For example, attempt to identify a security incident regarding suspicious activity originating from a specific IP. Threat information provided by external sources provide the values to the necessary attributes and by processing the LPA's logs and assets inventory an attempt will be made to identify the threat within the system.
- Attack pattern matching. By analysing network and system activity against attack patterns, potential security incidents can be identified.

It is worth emphasising that the patterns used for detection purposes can be either generic or organisation-specific. The first category includes more "universal" cases, such as a brute-force attack attempting to guess user passwords. The second category includes cases that are closely related to the organisation's business operations and/or information systems. For example, if its employees are expected to interact with the information system only within a specific time frame (e.g., 09:00–17:00), then a suitably crafted pattern could easily detect any user activity outside the designated time frame. Similarly, a significantly high number of reads by the same user for a given database may mean that the said user account has been compromised and data is being downloaded/extracted from the database. Provided that the database system has transaction logging enabled, a suitable pattern would be able to detect the aforementioned behaviour.

The establishment of these baselines and thresholds is also one of the outcomes of the system analysis phase that is conducted by applying the SSM. The figures representing each organisation's behavioural profile under normal circumstances are expected to vary, as they are dependent on both the size and the type of the organisation. However, during the pilot phase of CS-AWARE which primarily targets LPAs, the baselines and thresholds are expected to be in proportion to the organisation's size. For instance, a larger LPA that serves more citizens is expected to have a higher daily average number of database read/writes than a smaller LPA.

What is more, certain patterns are provided in a generic form to avoid bearing too much detail. Nevertheless, they still need to be adopted according to the structure and set-up of the organisation's information system, in order to function properly and be effective. For instance, detection of the failed user log-in attempts in a custom software application may require examining specific database tables and columns that have non-generic names.

The threat detection engine also receives cyberthreat intelligence (CTI) data and incorporates it in the analysis it performs to improve threat detection. For example, assuming that CTI provides a list of IP addresses that have been designated as being malicious, the threat detection engine will also perform a search for these IP addresses within the gathered log data. Should matches be detected against certain rules, an alert will be raised so that the system administrator or a cybersecurity expert will be signalled to further look into the matter. An example of such a case could be when certain malware exists within the organisation's information system and is trying to contact a command centre (whose IP address has been blacklisted as malicious).

4.2.1.4 Multi-Lingual Semantics Support

This component essentially consists of two subcomponents, each performing a different function: The NLP Information Extraction component and the Multilanguage Support component. The former subcomponent attempts to extract any cybersecurity-related information from social network feeds (e.g. Twitter) that could be proved useful for the CS-AWARE threat detection engine. This approach was based on the fact that cybersecurity experts and companies tend to share cybersecurity-related information through social networks (e.g. a new malware kind exploiting specific ports or services), well before notifying the official repositories, such as the Common Vulnerabilities and Exposures repository. Another example could be that of using the published information proactively. For instance, given the sentence "Google releases security updates for Chrome", it will be processed by the NLP component and the information it contains can be forwarded to system administrators in almost real time, thus informing them that they could/should schedule an update of the said software.

Any alerts and information about cyberthreats that exist in a language other than the system administrator's mother tongue can impose an extra layer of complexity, while trying to deal with them. The Multilanguage Support subcomponent thus aims at translating any kind of text written in a foreign language to the end user's mother tongue.

4.2.1.5 Data Visualisation

The data visualisation component is responsible for visualising the threats, the threat level, the possible self-healing strategies and the information shared with the Cybersecurity community. Moreover, it communicates back to the system any user responses related to the aforementioned functions, as well as lower-level administration. Since it is the only system component that conveys cybersecurity awareness to the user, it is presented in more detail in Sect. 4.3 below.

4.2.1.6 Cybersecurity Information Exchange

In the cybersecurity field, any information or knowledge about a threat (newly-discovered or not) is very valuable for proactively protecting organisations and for detecting threats more efficiently and effectively. For instance, if the characteristics of a threat discovered in an organisation are made public (e.g. which ports it attacks or uses to propagate, IP addresses it originates from), then another organisation can do the following:

- Proactively block the malicious IP addresses to protect itself from being attacked in the first place.
- Perform a targeted search in its log files to determine whether traces of activity resembling the behaviour of this given threat also exist in their system and, nevertheless, they have not been detected yet by the installed security mechanisms.

The cybersecurity information exchange (CIE) component is responsible for the transmission of cyber-threat intelligence (CTI) to external entities. CTI is made available in a structured form and more specifically according to the STIX 2.x standard, to facilitate interoperability and consumption of the provided information by external entities, such as threat intelligence platforms, Computer Emergency Response Teams (CERTs) and Computer Security Incident Response Teams (CSIRTs).

However, before CTI can be shared with external entities, there are several interoperability and security issues that have to be dealt with and they fall within the following categories:

- **Legal.** Restrictions on what to share (e.g. for reasons of data privacy) and perhaps obligations and/or liabilities to share.
- **Policy.** Existence of a suitable business policy specifying, among others, what to share, who to share it with and when.
- **Technical.** Semantic issues relating to how CTI can best be described and the suitable means for sharing (e.g. data types, formats, protocols, standards).

For complying with the legal requirements, sharing of CTI must be controlled. In order to protect privacy and personally identifiable information (PII), data anonymisation should take place, if necessary, prior to sharing CTI with external entities. An organisation's policy should contain well-established and clearly defined procedures, as well as appropriate deployed measures to prevent the leakage of classified or sensitive information. To this direction, data sanitisation can be employed to ensure that classified or sensitive information is not disclosed to external entities through the CTI sharing process.

4.2.1.7 Self-Healing

Self-Healing aims to assist LPA administrators respond to identified vulnerabilities and high-risk threats by providing customised healing solutions or recommendations. The Self-Healing component is a fully-supervised solution that uses the results of the Data Analysis component and looks for the most appropriate mitigation solution among its self-healing database and those provided by the external sources. The chosen solution, according to the administrator preferences, will either be applied automatically or request their approval or simply be presented in the form of a recommendation.

The CS-AWARE self-healing is an innovative, fully supervised system responsible for matching the results received from the Data Analysis component with mitigation rules stored in the Self-Healing Policies database to compose and propose or apply the most appropriate mitigation rule(s) for a given threat. The composed rule aims to mitigate the threat that was identified in the LPA's system. Furthermore, the CS-AWARE Self-Healing component has the ability to autonomously diagnose and mitigate threats, while ensuring that the system's administrator who is always aware of the system behaviour, can prevent configuration changes that may raise incompatibility issues.

Once the Self-Healing receives input data from the Data Analysis component it identifies the threat type and composes the proper mitigation rule autonomously. The composed rule is then sent to the Visualisation component for the administrator to approve or decline the application of the composed mitigation rule. The Self-Healing component utilises CTI data obtained from external sources and which includes indicators, security alerts, techniques and reports. This information facilitates the mitigation of threats which may have not been previously detected in the system.

4.2.2   The CS-AWARE Framework

The various components comprising the CS-AWARE solution were briefly presented in Sect. 4.2.1. A high-level view of the information flows among these components is presented in Fig. 4.1. Furthermore, the components have been divided into three distinct layers:

- The Data Extraction layer covers all components responsible for defining relevant data and extracting it, as well as the sources themselves.
- The Data Transformation layer summates all components tasked with transforming and analysing the data in some way. The data may be filtered and adapted, if necessary, before being processed by the modules.
- The Data Provisioning layer is meant for visualising and sharing the detected incidents and data patterns, as well as for the automated system reactions and recommendations performed by the self-healing module.

4.3   Development and Integration

The CS-AWARE modules can be divided into two main categories: The ones that were developed from scratch and the ones that were based on already existing components. The latter had to be adopted accordingly, in order to offer the required functionality. In this Section the adaptation of the existing components is presented, followed by the various decisions that had to be made and/or taken during the integration phase, as well as any challenges encountered.

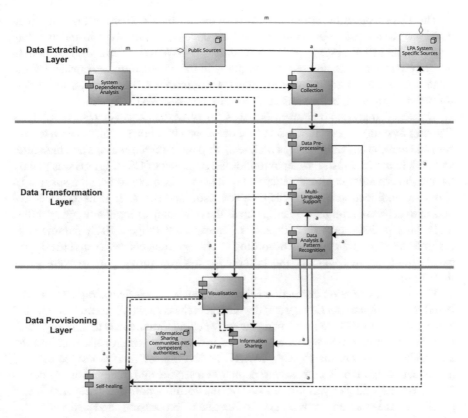

Fig. 4.1 Information flows within the CS-AWARE framework

4.3.1   Existing Components Adaptation

Some of the components comprising the CS-AWARE solution were developed from scratch, whereas others were based on existing technologies that were adopted appropriately, according to the requirements.

The System and Dependency Analysis Support Tool is based on GraphingWiki, an extension of the open-source wiki engine MoinMoin that was developed in 2006 to support the collection and visualisation of data. It was successfully used in software protocol and malware analysis by introducing additional meta link syntax and (other capabilities). The tool was further adopted to meet the needs of CS-AWARE by creating a new interface and new graphical elements for visualising the relationships among the nodes (e.g. double-headed arrows). Its functionality was further extended to support exporting of the contained information in JSON format, to facilitate the automated processing of any modules requiring access to it.

The Data Collection component was based on data collectors that had been developed in the past. For acquiring the logs data, collectors were installed at key points of the LPAs' systems and they feed the CS-AWARE system with the data to be analysed for threats. In order to comply with the GDPR requirements and at the same time keep the complexity low, data anonymisation was decided to be performed at the source, before transmitting it to the CS-AWARE system.

The Data Analysis component is based on MAARS (Multi-Attribute Analysis Ranking System), a proprietary software developed by Peracton that is able to process and analyse any number of parameters, with their unique settings and then filter and rank items very fast. Its adaptation for the purposes of CS-AWARE was mainly the creation of suitable threat patterns. For instance, in order to detect a suspicious database modification attempt, the database audit and access logs need to be processed to detect several parameters, such as the frequency of login attempts per day, login time periods, and IP addresses of unexpected ranges. Each parameter is assigned an appropriate weight according to its importance/severity and the aggregated result denotes whether the criteria for a given threat pattern were satisfied or not.

The NLP Information Extraction subcomponent was based on Graphene, a rule-based information extraction system developed in the context of research conducted at the University of Passau. The main strategy behind Graphene is to simplify complex sentences before applying a set of tailored rules to transform a text into the knowledge graph. During the CS-AWARE project the research prototype evolved into a technology that is both easy to deploy as a service and integrate as a product. A new extraction layer was also added for transforming complex categories into a graph of fine-grained knowledge, to facilitate processing and extraction of conclusions.

4.3.2 Planning the Components' Integration

CS-AWARE was originally designed to be installed as a cloud-based architecture, using serverless components and other features that are not available in traditional internal datacenter setups. However, during the pilots implementation phase, the original plan had to be changed due to an inflexible requirement of deploying the solution only inside a municipal datacenter (Roma Capitale).

At the same time, development was being performed by separate groups and it therefore became evident that the development process would benefit from using a separate virtualisation/container engine for each component.

The decision to use Docker as the component instance and Docker Compose as the integration framework was taken quite early. The main reason that steered the decision towards this direction was that this approach had been followed by the University of Vienna in the past and it seemed very promising, as it could deliver both local and cloud-implementations with relatively little effort. This architecture has proven its value so far.

When choosing Docker, it is also possible to use container-orchestration tools like Kubernetes, also part of Docker Enterprise. However, for the purposes of CS-AWARE, it was assessed that it would add more complexity than benefits to the project. Nevertheless, since Docker was being used anyway, it was possible to add the solution's custom setup to existing Docker Enterprise/Kubernetes server farms with relative ease.

When architecting applications like CS-AWARE, scaling is an important factor that needs to be considered. By using Docker it was possible to have a full-scale setup running on a well-equipped laptop. On the one hand, this gave the ability to perform development in a production-like environment. On the other hand, for the actual production environment, it will be possible to split the modules onto high performance servers, as well as split database and storage with ease during configuration. In this way, the load could be handled very efficiently and the aforementioned flexibility proved to be very valuable in the development and deployment process.

4.3.3   Integration Challenges

Due to the nature of the log data that would be collected for analysis, there were serious privacy issues that had to be dealt with, since, among others, IP addresses and usernames were anticipated to be contained in the municipalities' system logs. Significant effort was put in ensuring that there would be no privacy violations within the scope of the CS-AWARE project. This involved communicating with the competent Authorities of Greece and Italy, since the pilots would take place in the municipalities of Larissa and Rome, respectively. Furthermore, GDPR came into force on May 25, 2018 and consequently the processes of data collection, storage and processing were reviewed to ensure compliance with the provisions of the Regulation.

Another issue that primarily affected the self-healing functionality was the reluctance of the pilot LPAs to give extensive access to their systems. This was mainly due to security-related regulations and liabilities they were bound to, nevertheless, the human nature itself also seemed to affect it to a certain extent. Namely, despite the fact that the self-healing module could be configured to interact directly with a wide range of systems and thus offer automated reactions to the discovered threats (to the best possible extent), the pilot LPAs' system administrators seemed to want more control over anything that would happen to their system. This led to modifications in the user interface, to introduce the functionality of approving the suggested actions before they were actually launched. As a consequence, the self-healing functionality would operate in a supervised manner.

4.4   Interface Design for Increased Awareness

As one of the main features of the project was to convey cybersecurity awareness, and as this was not build on existing solutions, a novel approach was taken with innovative visualisations and interfaces to the different components.

In this Section a detailed account is given about the evolution of the user interface, starting from the initial ideas and thoughts about how it should look like for conveying awareness to the user, on to the various transformations it underwent according to the user feedback that was received in several iterations.

4.4.1   Initial Thoughts on Conveying Cybersecurity Awareness to the User

During the first workshops with the participating municipalities, it became evident that for the end users the visualisation interface should feature a consistent and simple overview that at the same time could convey a high level of information.

The cybersecurity awareness status in most organisations is lacking, as we quickly found out. The situation in larger organisations, like Roma Capitale, we established, is that there are already numerous cybersecurity tools, that show a lot of data, usually only what has been resolved by the systems themselves and only very narrowly focusing on the particular vendors equipment (a vendor usually provides only tools for their proprietary equipment, not for any competitors equipment) or only parts of the infrastructure (Personal computers as opposed to servers and network equipment). Smaller organisations like Larissa may be hampered by lack of common overview and may only be partially covered.

A system manager from Roma Capitale showed us a typical system from a hardware vendor, which proudly showed how many and how many kinds of incidents, their cybersecurity system had avoided in those particular servers of this vendor, spanning several pages that you could browse. His comment was reluctant: "but what can I use it for?" all cases shown were the resolved cases, the overview was miniscule and it only cared for this particular vendor. In other words, the current tools did not convey the needed level of cybersecurity awareness.

In Larissa, anecdotal evidence for lack of overview and cybersecurity awareness, was the description of how they once understood that there was an ongoing DDOS-attack, where a perpetrator makes thousands of (typically) infected PCs send request to a particular service in order to try to eliminate a particular service of the municipality. They did not have a common system to show what system the attack was targeting, but could hear that the fans in the server room next door started to work at high speeds - and hence noisy, so they realised that something was going on. Had they been out of office, working from home, as they had to in the times of the pandemic, they would never have heard the noise and would have been even slower in detecting the cybersecurity incident. The antivirus system that was used does not show this kind of attack, and is hence lacking in conveying cybersecurity awareness.

We already covered the detection and handling in the analysis module earlier, but the visualisation giving the overview and reason for a system administrator to want to use a system, was a real innovation to be done: a different approach was needed to increase the situational awareness in both smaller and larger municipalities, thus satisfying the primary requirement of the project.

As the system will be used internationally, it was important to add language support for two purposes: first of all, all menus and commands can be changed into the language of the user and secondly, using either external or internal translation services, the messages from cybersecurity organisations (e.g., CVE's) or social media, are translated automatically.

In order to cater for security in the interface, a role-based login-system was developed that was easily configurable for a skilled developer. This enabled a fast integration into the proprietary system of Roma Capitale.

4.4.2   The Evolution of the Interface According to User Feedback

From the outset, it was important for the visualisation team to find metaphor that would engage the users. This was done from a cognitive philosophical angle, as metaphors convey much stronger value of the visualisation if they give meaning to the user as described in e.g. Metaphors we live by, (Lakoff & Johnson, 2008) and Women, Fire and Dangerous Things (Lakoff, 1987). In The Contemporary Theory of Metaphor (Lakoff, 1992), George Lakoff writes:

"Metaphor allows us to understand a relatively abstract or inherently unstructured subject matter in terms of a more concrete, or at least a more highly structured subject matter." (Lakoff, 1992, p. 245).

This leads to the choice of the dartboard visual that gives the immediate overview of the threat situation of a municipality and raises the awareness of the user, as the picture of a dartboard automatically raises the attention of the user, and hence the awareness of the issues at hand.

Using a dartboard metaphor, the kind of dartboard with divisions, the arches came to be used to divide the threats into groups, and in our case, different cyber threat groups, as well as showing the number of threats by their extent and their criticality using colour. To make it easier to follow trends, it would also incorporate visualisation changing over time. All the aforementioned functionality was chosen in order to raise awareness of the cyber security threat level.

For visualising threat levels, it was also early decided to use common colour scheme, so that threat awareness would have a common ground and would be intuitively understood by the end user. Again, leaning back to the understanding of colour gained from Lakoff (1987), meant that the colours of the different threat levels needed to be significantly different from each other in order not to confuse the user. For a fancy graph in a newspaper, you may choose colours with limited variation in hue, but as Lakoff shows, some cultures do simply not see the nuances as other cultures do, so we determined that it would be best to have significant colour differences. The initial colour schema only consisted in three colours, green, yellow, red.

The initial concept of the user interface emerged quite early in the process, as the visualisation team from experience knew that visual progress needed to be done early to get the same picture in the head of all participants as early as possible, in order to avoid working in a number of unknown different directions.

The very first draft of a potential visualisation scheme was presented in the Vienna consortium meeting, in February 2018 (Fig. 4.2).

Here we see the first implementation of the ideas above, visualising the threat levels: a circular graph similar to a dartboard was used, where the threats were shown in groups and their current prevalence was depicted via the size of the arch they expand upon. In this way, any number of threat groups could be shown, and it would be easy for a system administrator to pick out the most dangerous threats, to see which should be dealt with first. The "bull's eye" in the centre of the disk indicates the overall risk level. When clicking on an item (a threat group) in the dartboard it was envisaged that one or more modal windows would open and provide more details about the individual threats that had been identified, the way to resolve it and external intelligence sources. This design was meant for making it simple for the system administrators to handle the threats discovered by the CS-AWARE system. Notice the three colours of the arches.

After good response from the meeting, both from the pilot project municipalities and from the other partners, the colour scheme was enlarged sot that it would match the colours of the threat levels of the United States Department of Homeland Security using the colours green, blue, yellow, orange and red in increasing order of

Fig. 4.2 The first draft of
the visualisation scheme
(February 2018)

severity, as we determined that the awareness would benefit the most from this scheme (Fig. 4.3).

A few months later, in the Thessaloniki meeting (October 2018), the first functioning graphical interface was presented to the consortium members (Fig. 4.4).

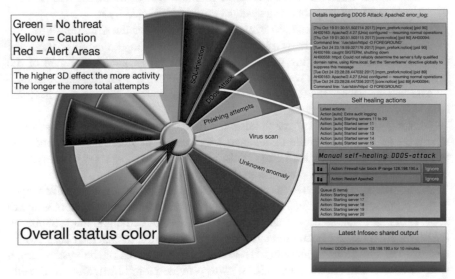

Fig. 4.3 The 5-scale severity levels

Fig. 4.4 The first graphical version of the visualisation (October 2018)

After development of the modules and at the preliminary stages of the integration phase, it was possible to try the first version of the CS-AWARE system, where all components could successfully communicate among themselves. Feedback from both the consortium members and users, indicated the need for exploiting the system dependency overview captured by the Soft Systems Methodology workshops (that took place in the municipalities) and stored in the GraphingWiki tool. Using this information would enable the system to show where the threat was initially discovered and what neighbouring systems could be affected by it.

Instead of building a separate instance in the setup for the GraphingWiki, the visual interface was further developed and enhanced with the same content and tools for building and editing a network, so it now also contains the map of the municipality networks that were captured at the workshops with all stakeholders.

The system reached the following look by the mid-term review meeting in Rome (Fig. 4.5).

Clicking on the "System" menu button, would open the system graph window, where the administrator could understand very quickly at which network node the threat was detected and which ones could also be affected by it (Fig. 4.6), as the nearest - in network terms - nodes were highlighted.

Additional user input led to the expansion of the workflow engine, small changes in the colour scheme (for consistency reasons), enhancement of the system graph to show different symbols for different types of entities and changed the grouping in the dartboard overview from threat groups to IT categories, to provide better awareness for the system administrators. The treats and the nodes of the system graph were enriched with additional detail, so as to provide more information to the end users and help them understand the problem better. In order to care for workflow, a workflow engine was also developed, so that the primary System Administrator/Investigator could send the threat to a particular expert, e.g. a database expert or a network expert, depending on the nature of the threat.

Finally, search functionality was added to the system graph and keywords were added in the social media overview.

Fig. 4.5 The CS-AWARE interface by the mid-term review meeting (April 2019)

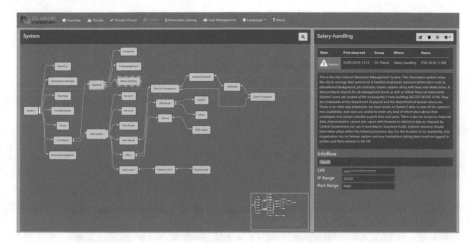

Fig. 4.6 Visualisation with incorporated system graph, showing the various system nodes and their interconnections that could also be affected by the same threat

4.5   Conclusions

This chapter presented the various stages during the development and implementation of the CS-AWARE solution, a system meant for raising cybersecurity awareness. For raising awareness in an effective and efficient manner, significant effort was put on threat contextualisation, so that the detected threats were correlated and filtered according to the particular qualities of the monitored organisation. Through the application of the Soft Systems Methodology, a detailed mapping of all nodes worth monitoring was created and this information was made available to other modules so that they were able to produce more targeted outputs, with the common aim of raising awareness as much as possible.

Each of the various components comprising the CS-AWARE system plays a particular role in raising the awareness of the end user, who is expected to be an IT system administrator. More specifically, the data analysis component has the ability to combine input from multiple sources (log files from the organisation's nodes, external cyberthreat intelligence, posts on social networks), analyse it and show what is happening (via the visualisation component). In this way, the cybersecurity awareness of the system administrator is increased, as they gain insight about cyberthreats that have affected or could affect their systems. Hence, the analysis results are very refined, since both the importance of a given node and the severity of a threat has been taken into account.

The cybersecurity information exchange component also helps in raising cybersecurity awareness, by gathering cyberthreat intelligence from a world-wide spectrum. Presenting this knowledge to a system administrator can show them what the latest trends are, how sophisticated malware can act and spread, and so on. This knowledge itself can be an eye-opener, since it can help administrators to think like attackers or malware writers, which in turn can lead to improved procedures and measures that better protect the information system.

Finally, via the multilingual semantics support component, the language barrier was eliminated, thus enabling system administrators to understand better the outputs of the CS-AWARE system (both textual and graphical). Hence, this minimised the possibility of misunderstanding something or not paying the necessary attention to a detected threat.

References

Lakoff, G. (1987). *Women, fire, and dangerous things: What categories reveal about the mind.* University of Chicago Press.
Lakoff, G. (1992). The contemporary theory of metaphor. In A. Ortony (Ed.), *Metaphor and thought (pp. 1–50).* Cambridge University Press.
Lakoff, G., & Johnson, M. (2008). *Metaphors we live by.* University of Chicago Press.

Chapter 5
Deployment and Validation of the CS-AWARE Solution at Two Pilot Sites: A Combined Agile Software Development and Design-Based Research Approach

Jerry Andriessen, Thomas Schaberreiter, Christopher Wills, and Kim Gammelgaard

5.1   Introduction

Deployment can be defined as locally implementing technology and making sure that it functions in the designated contexts. The technology has been designed and is functioning properly, in a technical sense, and will now be deployed and tested in actual contexts, in a socio-technical sense. In this chapter we report how we deployed and validated the CS-AWARE solution in two municipal contexts.

We will use the term validation, rather than evaluation, because this is technically more correct. In the domain of technology, validation is defined by Wieringa (2014) as the assessment of a simulation of the technology in a simulation of its intended context of use, in order to predict what would happen if the technology were actually used by stakeholders in this intended context. This does not cover exactly what we will be doing, as we are working with an actual implementation of the technology in the intended context, but we do work with simulated scenarios emulating cybersecurity attacks. An evaluation study would assess what has

J. Andriessen
Wise & Munro, The Hague, The Netherlands

T. Schaberreiter (✉)
CS-AWARE Corporation, Tallinn, Estonia
e-mail: thomas.schaberreiter@cs-aware.com

C. Wills
Caris Partnership, Fowey, UK
e-mail: chris.wills@cs-aware.com

K. Gammelgaard
RheaSoft, Aarhus, Denmark
e-mail: Kim@rheasoft.dk

© The Author(s), under exclusive license to Springer Nature Switzerland AG 2022
J. Andriessen et al. (eds.), *Cybersecurity Awareness*, Advances in Information
Security 88, https://doi.org/10.1007/978-3-031-04227-0_5

happened in the actual use of the technology after it has been transferred in practice, which, unfortunately, was beyond the time-scope of the project. In social sciences, the term formative evaluation is often used in the same sense as validation in technology through development studies (e.g Thomas et al., 2019). Therefore, in this chapter, when we use the terms validation and evaluation, we mean the same thing: formative evaluation in the context of technology deployment. We want to underline that our approach subscribes to a systemic vision that considers the technical solution as part of a more complex social system in which people live and interact (Piccolo & Pereira, 2019). The notion *cybersecurity awareness* refers to this systemic vision, where we suppose users will find themselves, on a daily basis, needing to make security-related decisions, and they clearly need a level of awareness and understanding in order to do so (Furnell & Vasileiou, 2019).

We will not discuss details of technical deployment in this chapter, our focus is socio-technical. This means that we will report about deploying and validating the technology in the authentic user contexts and we will report on involvement and feedback from users in all phases of deployment. We started with collecting user expectations at the start of deployment. Then, components of the (implemented) technology were tested, user feedback was collected and interpreted, and modifications were implemented, during three cycles of 3 months. For each cycle, we implemented user feedback as an ongoing process until the very end of the project, this collaborative process is called agile software development (Strode, 2016). The cyclical approach of evaluation, revision and evaluation is called design-based research (Barab & Squire, 2004; Berliner, 2002; The Design-Based Research Collective, 2003). In our approach, these two go very well together, even though from the social sciences point of view, where we speak about formative evaluation, from the computer-science point of view, this is called validation.

For monitoring and validation, the main input came from frequent usability testing and from regular interactions with users from pilot deployment teams formed in each municipality. Through these qualitative methods, we gathered valuable information about how the system was used, and were able to reach higher levels of refinement of its functionalities. In combination with the agile approach, feedback could be implemented quickly on the technical level. In addition, through the use of questionnaires, we got informed about how users perceived the system, in terms of general characteristics, usability, their awareness of cybersecurity, and the impact on the organisation.

This chapter provides a stepwise walkthrough the deployment and evaluation activities of the CS-AWARE cybersecurity awareness solution at the municipality of Roma Capitale, and at the municipality of Larissa, during three cycles of design, collecting feedback, testing and revision (see Table 5.1 below).

In this chapter, we (in Sect. 5.2) briefly summarise the activities that involved users in building the CS-AWARE system. We then (in Sect. 5.3) present the deployment scenario, and our evaluation methodology (Sect. 5.4). In subsequent sections, deployment and evaluation outcomes for each of the cycles are described, summarised and main conclusions will be derived. In our conclusions, we will also

Table 5.1 Three Deployment cycles

Cycle	Period	Months in project	Section in this chapter
1	Oct-Dec 2019	M26–28	5.5
2	Jan-March 2020	M29–31	5.6
3	April–June 2020	M32–34	5.7

focus on differences between the two municipalities, the lessons learnt, and on methodological issues. Final conclusions are presented in Sect. 5.8.

5.2 Establishing the Socio-Technical Context

The main objective of CS-AWARE (https://cs-aware.com/) is to provide a socio-technical cybersecurity situational awareness software solution for Local Public Administrations (LPAs), that will in principle work for any small- to medium-sized enterprise (SME) or large organisation. This solution enables the detection, classification and visualisation of cybersecurity incidents in real-time, supporting the prevention or mitigation of cyber-attacks. It is intended that this solution will provide an intermediary step towards automation of cyber incident detection, classification and visualisation, that impacts on cybersecurity awareness of users. User involvement is crucial in all steps of design, implementation, deployment and evaluation. In this section, we briefly revisit (see Chap. 2) the three iterations of workshops where we captured essential information for building the application during the first 25 months of the project. We do this because these workshops have strong user involvement and therefore contribute to awareness of cybersecurity in the LPAs. After this period, we distinguish three deployment cycles, each of about 3 months, during which we systematically tested and revised the CS-AWARE application in the context of the two pilots.

The method exploited in three workshops is part of the analysis according to the soft systems methodology (SSM, Chap. 2) that was chosen for CS-AWARE. In the SSM approach the analysis and identification of assets, dependencies and monitoring points of the existing and organically grown complex socio-technological systems found in all larger organizations—like Local Public Administrations (LPAs)—is an integral part of the proposed cybersecurity awareness solution. We argue that in complex systems good cybersecurity awareness can only be provided if the relevant relations between the mission critical aspects of the system are understood, and relevant case specific monitoring points can be utilized.

SSM-Workshop 1: The first workshop focused on getting an overview analysis of cybersecurity related aspects, relevant to CS-AWARE in the context of LPAs. The analysis is based on three thematic focus points: an initial threat assessment for LPAs, an analysis of external information sources that may be relevant to the solution, and an analysis of the pilot scenarios (relevant work flows) collected from the users during the first round of workshops. We have investigated potential monitoring points at four different levels that allow to identify suspicious behaviour related to data operations: the database level, the application/service level, the network level and the security appliance level.

SSM-Workshop 2: A second workshop was organized to refine our understanding in the three main thematic focus points covered in the first iteration, as well as to assess a fourth focus point, the definition of CS-AWARE use cases, based on our understanding and results of the first three topics. This allowed us to identify the critical processes of the services that are used for CS-AWARE piloting, and the associated information flows through the system that those processes create in day-to-day operations. To achieve this understanding, the workshops were organized in two parts: the system and dependency focused workshop to refine the understanding of systems and processes already started in the first round of workshops, as well as a more end-user focused story telling workshop to determine the cybersecurity related problems users and administrators alike face on a daily basis, and the procedures and processes used to solve those problems. This second workshop is fully described in Chap. 3, and the implications for awareness are taken up again in Sect. 5.7.

Four specific use cases could be identified:

1. Vulnerability awareness: map classified vulnerabilities to specific LPA systems/components.
2. Behaviour analysis: identify suspicious behaviour and if possible, classify according to data received from threat intelligence.
3. General security warnings: informing about general and/or currently ongoing security events that may become relevant to the specific context of each LPA
4. An analysis of connections originating from or going to specific IP/DNS entries that are classified as malicious by relevant communities.

We assume that the list of use case scenarios covers all aspects relating to cybersecurity awareness that can be covered considering the data provided by the various communities (see also Chap. 1 for discussion of cybersecurity threats).

SSM-Workshop 3: In the third workshop the goal was to define, together with the LPA users and building upon the first two iterations, normal and abnormal behaviour within the identified business processes (pilot scenarios), and how this behaviour is reflected within the data sources collected from the LPA systems on the database, service, security appliance and network level. The behaviour of system elements during day-to-day operations according to the identified business processes, and how this reflects in the data sources CS-AWARE collects, is a crucial input for the definition of accurate and relevant monitoring patterns. The resulting patterns were validated through the consent of CS-AWARE security and data analysis experts, as well as the employees of the Municipalities (users, administrators, managers) who are the ones the cybersecurity awareness system is intended for. Similarly, self-healing policies have been defined that allow mitigation of events detected by cybersecurity patterns in an automated way.

During the first 25 months of the project, we developed the system, whereby we implemented the outcomes of the workshops. Then, deployment and evaluation started.

5.3 The Deployment Scenario

For better understanding of the socio-technical aspects of deployment of the system in the municipal contexts, we developed a systematic co-creation approach, called a deployment scenario (Fig. 5.1). Similar to the joint efforts in the SSM- workshops, co-creation of the elements of a deployment scenario (by researchers and stakeholders at the municipalities) can be seen as part of a trajectory towards more awareness of cybersecurity in public administrations (Schønheyder & Nordby, 2018). This makes deployment also a learning trajectory, with greater awareness of cybersecurity as the outcome. Our approach is similar to what is called project articulation, which starts with developing a joint vision rather than with a list of things to do (Strauss, 1988). To support our municipalities in developing clear goals and expectations, and in articulating these together, we employed a scenario template, to guide the discussion with the pilots. This template was based on earlier experiences with this procedure in another project (Ambrosino et al., 2018).

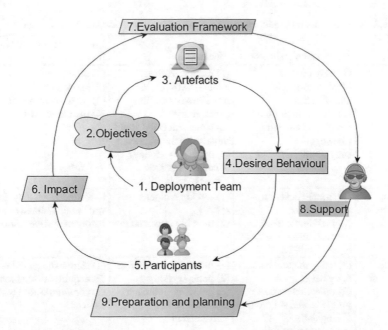

Fig. 5.1 The Deployment Scenario

The template was applied during the same meeting as the third SSM workshop. We addressed and discussed each scenario element with the users. The template was revisited with the same users after 1 year, but they did not feel the need to make any changes until the end of deployment.

The nine elements of the scenario are as follows:

1. The Deployment Team: At the pilot sites, a local deployment team is established. The role of the deployment team is to design, prepare and monitor execution of the deployment scenarios in coherence with the requirements. In Rome, the deployment team consisted of system administrators (n = 6), managers (n = 5), and local users from the municipality (n = 2). In Larissa, the team comprised a manager/owner, 3 system administrators and their unit manager, and two service users from the municipality. Members of the deployment team are also involved in validation activities.
2. The Deployment Objectives: These are negotiated local objectives for deploying the system, made explicit by the different stakeholders. The objectives for each municipality can be seen in Tables 5.2 and 5.3.
3. Desired Artefacts: This is the tangible output from the CS-AWARE system that users desire for their context. This means the concrete output of the system: a report about incidents, an overview of what was shared, what healing operations succeeded, etc. The outcomes for each municipality can be seen in Tables 5.2 and 5.3. Chap. 8 will further elaborate on user-role specific interface options.

Table 5.2 Main negotiated instantiations of the deployment scenario for Roma Capitale

	System administrators	Managers	Local users involved in services
Objectives	(1) easy treat identification and classification (2) reduction of time for threat understanding (3) mitigation suggestions	(1) more effective relation with service providers in handling cybersecurity (2) better informed about threats and mitigation (3) quality of services (4) more satisfied citizens	(1) efficiency of work (2) service reliability (3) personal data protection (4) increased trust
System artefacts	(1) trouble ticket on mobile device (2) real time info on number, resolution time, types of attacks	(1) weekly incident reports (2) monthly trend reports	Users indicated they did not want to be notified of threats before they were resolved
Desired behaviour	(1) using the tool on a daily basis (2) reflection on past problems and solutions (3) regular communication with technical team and internal users	(1) improved knowledge (2) policy making (3) proactive approach to users and senior management	(1) follow the guidelines (2) acquire information (3) communication with citizens

Table 5.3 Main negotiated instantiations of the deployment scenario for Larissa

	System administrators	Managers	Local users involved in services
Objectives	(1) timely detection of treats (2) no false alarms (3) up-to-date information (4) trustworthy information (5) room for making your own decisions	(1) no service down time (2) no extra costs or resources (3) no reputation damages	(1) no additional burden (2) feel safe and protected (3) timely, clear, concise information (4) not being watched
System artefacts	(1) ranking of threats by frequency (monthly) (2) list of system affected nodes (weekly) (3) report of information shared by other LPA's (monthly)	(1) weekly incident reports (2) weekly report of services affected	Weekly report of threat sources
Desired behaviour	(1) assigning monitoring roles on a daily basis (2) actively learn from the system (3) collaboratively discussing solutions	(1) Reading the reports (2) increased trust in system administration	Better following the guidelines

4. Desired User Behaviour: What will users do to make sure that the objectives will be realised? For awareness it is important to understand: what effective (and less effective) stakeholder activity looks like. How can we observe things are going well, as desired, or less so? The outcomes for each municipality can be seen in Tables 5.2 and 5.3.
5. Participants: we discussed within the deployment team who will participate in the various meetings and tests in using the system.
6. Impact: Impact of deployment is what users expect that will change in the organisation when deployment objectives will be realised, in a long-term perspective. Thinking about desired impact allows us to understand 'how far we are'—what can be realistically achieved, in the eyes of the users.
7. Evaluation: Evaluation will be carried out at four levels, corresponding to the four main research questions (Sect. 5.1): (a) The technical level of system functioning; (b) Usability of the interface; (c) User awareness; and, (d) Organisational awareness. Evaluation methods are further explained in Sect. 5.4.
8. Support: Intended here is the support foreseen during deployment, within the organisation and/or by the deployment team: is there someone watching over the users, technical support, are there manuals or training?
9. Preparation and planning: this involved all preparatory activities for the deployment: technical preparation, implementation and testing, recruiting users, communication with users beforehand, introductions, manuals, preparatory meetings.

Tables 5.2 and 5.3 describe scenario elements 2, 3 and 4, for Rome and Larissa, respectively. After the deployment scenario was established, we implemented regular meetings with the local deployment team. During these meetings, we further discussed the status of the deployment scenario, and collected feedback on various aspects of deployment. Table 5.4 shows how we have exploited the scenario during deployment: for further implementation of the technology, for understanding and user feedback, and for evaluation.

5.4   Validation Methods

Formative evaluation of the CS-AWARE system is part of deployment. During deployment, aspects of this system were still being modified, as a result of internal discussions in the team, and of user feedback. Agile development of technology, combined with soft systems analysis during the workshops, does not enable the clear prediction of outcomes. Instead, what they afford is a solution that is closely tied to user needs and local circumstances. Therefore, in line with this approach to design, we developed an appropriate approach to evaluation, which is design-based, where evaluation is seen as a formative intervention (Abildgaard & Christensen, 2017; Berliner, 2002; Paavola et al., 2011; The Design-Based Research Collective, 2003). The following methodological principles for evaluation (quoted and adapted from Engeström, 2011) apply to our situation:

1. Participants face a problematic and contradictory object, i.e., cybersecurity, embedded in their vital life activity, which they analyse and expand by constructing a novel concept, cybersecurity awareness, the contents of which is not known ahead of time to participants nor to the researchers.
2. The contents and course of the intervention, using the CS-AWARE system, are subject to negotiation and the shape of the intervention is eventually up to the participants, who thereby gain agency (from awareness) and take charge of the process (ownership).

Table 5.4 Use of information from the deployment scenario

Deployment scenario element	Is used for…..
Team	Contacts & feedback
Objectives	Views on using the system Evaluation (c): User awareness
Artefacts	Information from CS-AWARE system
Behaviour	Evaluation (c-d): Awareness
Impact	Long term expectations Evaluation (d): Organisational awareness
Support	Immediate needs, manuals
Planning	Preparation and planning

3. The aim is to generate new concepts that may be used in other settings as frames for the design of locally appropriate new solutions.
4. The researcher aims at provoking and sustaining an expansive transformation process led and owned by the practitioners.

In practical terms, this means a cyclical approach: (a) we will be collecting evidence from the users: their ideas on cybersecurity and their activities with the CS-AWARE system; and (b) adapt the design (of one or more CS-AWARE system components) based on that evidence, and (c) revisit how this affects the ideas and activities of the users; etc. A researcher should therefore not impose understanding of the main concepts, it is our aim is to understand how the users appropriate those concepts and use the technology.

Because this project was financed and reviewed by the European Union, we decided it would be important to also collect quantitative evidence to evaluate the CS-AWARE system. To this end, we developed a hybrid approach to evaluation that involves two sources of evidence (A) Qualitative evidence, derived from (1) the deployment scenarios, (2) user verbalisation and actions during usability testing of the system, and (3) feedback from users collected at regular intervals during meetings; and (B) Quantitative evidence, collected through questionnaires, one for each level of evaluation, and administered twice during deployment.

The hybrid approach serves to provide answers to the following research questions:

1. Technical Level: To what extent is the technical implementation effective?
2. Usability Level: To what extent is the CS-AWARE system usable by foreseen target users?
3. Awareness Level: To what extent is the awareness of users affected by discussing and using the system during deployment?
4. Organisational Level: To what extent does using the CS-AWARE system have an impact on cybersecurity awareness at the organisational level?

Qualitative evidence is used for understanding and evaluating how users work with the system, as well as how their awareness (qualitative understanding) of cybersecurity evolves during deployment. Quantitative evidence provides evidence on the attainment of key performance indicators (KPIs), related to requirements for each level, as will be specified in the next section.

5.4.1 Requirements, KPIs, Questionnaires

Within the context of the CS-AWARE framework (see Chap. 4), we formulated two sets of requirements that the system is supposed to meet (Table 5.5). The first set were technical requirements, and the others refer to technical system components (see Chap. 4). These requirements are functional: they pertain to functionalities that the system should offer. The second set involves non-functional general system

Table 5.5 System Requirements (F: Functional; NF: Non-Functional; EU: End user based)

#	Requirements	F	NF	EU
Technical system requirements				
S1	Provide cybersecurity awareness	X		X
S2	Allow information sharing	X		X
S3	Enable system self-healing	X		X
S4	Enable data collection from internal LPA and external cyber security information sources	X		
S5	Allow pre-processing to bring data into a unified format	X		
S6	Enable data analysis by setting external and internal data into context	X		
S7	Ensure international usability of the system by providing multiple languages	X		
S8	Identifying relevant internal and external sources	X		
General system requirements				
S9	Usability (the usability of the CS-AWARE system, as determined by the end users)		X	X
S10	Compliance (compliance to LPA regulations, policies and procedures)		X	X
S11	Integratability (Integratability of CS-AWARE system into LPA work flows)		X	
S12	Open source (how much of the CS-AWARE components can be open sourced and how much is kept proprietary)		X	
S13	Internationalization (integration into different cultural and language contexts)		X	

requirements, which are requirements for the system as a whole. These general requirements (except S12, 'open source') are to be tested with end users. The three main system requirements (S1, S2, S3) are also subjected to end user evaluation. The degree to which our approach is supposed to meet the requirements is expressed in terms of key performance indicators (KPIs), which are scores on a scale that maps test results to specific requirements, in terms of the research questions. For defining KPIs in the next subsections, we take the name of the subscale, not the score (which always has to be 60% or higher). As Fig. 5.2 shows, the KPIs are coupled to the four research questions, and provide therefore important evidence for the success of the CS-AWARE system on the four different levels addressed by the research questions.

For each evaluation level, we developed a questionnaire to be answered by our users at the pilot sites. The questionnaire items allow users to indicate the degree to which they agree with a statement (Likert Scale 1–5, a well-accepted scientific approach (Eyvindson et al., 2015; Janhunen, 2012; McKennell, 1974)). All questionnaires start (or end) with a section where the participants can indicate their experience with CS-AWARE, and their role in their organisation. We have set a baseline score per KPI at 60% appreciation, as this reflects the accepted level of satisfaction. Figure 5.2 provides an overview of the relationship between system requirements, questionnaires, and KPI's. In the current section, we briefly explain how the requirements for each level are addressed.

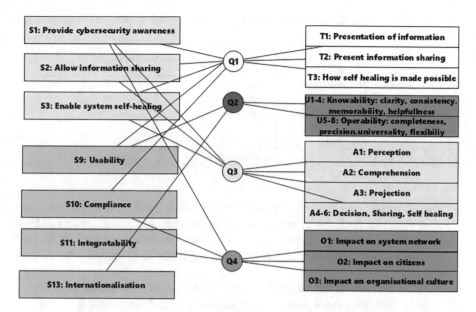

Fig. 5.2 System requirements (column on the left), questionnaire numbers, and KPIs (column on the right); Q1: System effectiveness; Q2: Usability; Q3: User awareness; Q4: Organisational impact

5.4.1.1 Level 1: Technical

Validation at this level addresses (1) technical validation and (2) the extent to which users assess the technology of CS-AWARE as sound (i.e.: it works as it is supposed to).

(1) The *technical validation* focuses on testing the functional requirements for each individual CS-AWARE component (see Chap. 4: system and dependency analysis—GraphingWiki, data collection and storage, data pre-processing, data analysis and pattern recognition, multi-language support, visualization, information sharing and self-healing), as well CS-AWARE integration testing. The component testing is based on the functional requirements defined for each component in Table 5.6. Some CS-AWARE system requirements (in particular the functional requirements, except S9-S11) are also evaluated based on technical criteria that do not involve users. The functional requirements mentioned are evaluated by requiring passing a set of the functional tests.

(2) *User validation* of the technology is done through Questionnaire 1. This questionnaire addresses the main requirements S1, S2, S3, S9, and S10. The KPI's that we link to these requirements for technical evaluation are the following:

Table 5.6 Overview of our hybrid evaluation approach during deployment, underlined methods are qualitative

	Cycle 1	Cycle 2	Cycle 3: Final Test
Technical level	Formal tests	Questionnaire 1	Questionnaire 1
Usability level	Usability test	Usability test, questionnaire 2	Usability test, questionnaire 2
User awareness level	Baseline setting Qualitative Analysis	Questionnaire 3	Questionnaire 3
Organisation level	=	Questionnaire 4	Questionnaire 4

- For S1, providing cybersecurity awareness, we inquire the extent to which the users (including new users, who have seen the system, as a demo) appreciate how information is presented in the various overviews and detailed views that the CS-AWARE system provides (see Chap. 4, and the sections on usability in the current chapter).
- For S2, provide information sharing, we evaluate the extent to which users appreciate the information selected from trusted sources, and the technical implementation at the level of the interface.
- For S3, system self-healing, we ask about user appreciation of the relevant information and the technical implementation of self-healing.
- For S9, usability, we have a general question about user appreciation. More detailed information is investigated in questionnaire 2, but only for users who have actually worked with the system.
- Finally, for S10, compliance, we ask about the extent to which the user thinks the system complies with internal regulations.

The 11 questions for level 1 are formulated in a general manner, e.g. 'I am happy with how information is presented in dartboard overview' (addresses S1), or 'I do not like how information is presented in system overview' (addresses S2), and 'I think the CS-AWARE technology complies with our regulations, policies and procedures' (addresses S10). The general formulation of the questions allows new users, i.e. users who have not worked with the technology, or only incidentally, and users from outside the LPAs, to fill in the questionnaire, for example after a demonstration session at a conference.

5.4.1.2   Level 2: Usability

Usability addresses requirement S9, that we will split up into more detailed requirements, and S13, the use of language features.

We adopted the taxonomy of usability provided by Alonso-Ríos et al. (2009). For our specific purposes, we formulated usability requirements for *Knowability* (user can understand, learn and remember how to use the system) and *Operability* (the system provides the user with the necessary functionalities). Both are split up into more detailed KPIs for usability. For Knowability:

- U1 (clarity) addresses the extent to which the elements on the screen (options, colours, used classifications) are clearly represented, easy to find, and have a clear function.
- U2 (consistency) addresses the extent to which the elements are consistently used throughout the application, and have a consistent structure and function.
- U3 (memorability) addresses the extent to which the elements are re-membered, including their structure and function.
- U4 (helpfulness) is about the extent to which documentation and online support-ive elements provide sufficient support.

For Operability, we have the following KPIs:

- U5 (completeness): addresses the degree to which the user thinks the information provided by the CS-AWARE system is complete, reliable, and helps in deciding what to do.
- U6 (precision): the extent to which the user thinks the information is what is needed to understand and resolve a specific threat
- U7 (universality): the extent to which the translations that are provided by the system are clear. This also addresses the requirement S13 (internationalisation).
- U8 (flexibility): the extent to which the system provides various types of users (system administrators, service provides, managers) the options they need for their role.

The 18 questions of Questionnaire 2 result in scores for each of these requirements for user evaluation of usability.

5.4.1.3 Level 3: Awareness

The specific requirements for awareness that we test are based on the phases described in a scenario for threat detection and mitigation by Hibshi et al. (2016): *Perception* (user perceives a threat), *Comprehension* (User understands the threat, its characteristics and information provided by the system), *Projection* (user fore-sees consequences of actions), and *Decision* (User makes a decision). To those, we add awareness of *sharing* and *self-healing*. The requirements for awareness link to the general requirements for cybersecurity awareness (S1), sharing (S2), and self-healing (S3). The requirements, combined with the phases, lead to a set of KPIs for awareness. For each of these, we indicated in what way the CS-Aware interface affords it:

- A1 (perception): which involves the affordances of the opening screen of the interface (Fig. 5.10): (a) detection of a threat; (b) estimation of its possible impact, on a dimension very harmful – innocent; and (c) table of main features of the threat: date, type, part of the system network involved. Users will now be immediately aware of a threat and its main characteristics. To what extent will users process and understand this information?
- A2 (comprehension): of the cybersecurity threat, a process for which the interface provides additional information about the threat, from reliable sources (Fig. 5.11). Please note, that the main goal of our approach is cybersecurity awareness, which goes beyond the resolution of a threat, therefore consultation of this information is a crucial part of this awareness requirement. The comprehension process (understanding the threat before action is undertaken) is supported at the console by the threat information screen. Providing necessary and useful information is a main research question for evaluation.
- A3 (projection): understanding future events or consequences, including potential, foreseeable attacks, or failures that result from poor security is a crucial aspect of cybersecurity awareness, which goes beyond detection and repair of threats. This awareness process is facilitated by the system and dependency graph (Fig. 5.14), which is an interactive visualisation of the system network, showing information about the network components when the user clicks on them, and also displaying the components that are linked, or functionally interdependent (i.e. being part of the same pilot scenario, such as salary administration). We ask users about their use of this visualisation, and the extent to which they are aware of the dangers of the threat to nodes in the system network.
- A4 (decision): involves the possibility of indicating resolution of the threat (resolution itself happens outside of CS-AWARE), applying self-healing (if this is available for this threat, and the user has the rights, it can be automatically applied), and also deciding to share the information about the threat with the relevant expert authorities or communities. There is some difference between the complexity of decision-making in Rome, where several people may be involved in threat resolution, and in Larissa, where one or two experts are sufficient. For this part of evaluation, we ask the users about their communications with others, and understanding (sometimes: being told) if a threat is resolved.
- A5 (sharing): here we ask users about the extent to which they understand the need to share information and about the local policies for sharing. We expect no high scores here until management policies have been considered.
- A6 (self-healing): for awareness, it does not only matter if the user can apply self-healing, more important is it that this happens with full understanding of the consequences.

Because awareness is a rich concept, Questionnaire 3 involves 36 questions. This questionnaire is applied in cycles 2 and 3.

For cycle 1, the phases that we described will also be leading in determining the baseline level of awareness, to be able to see if awareness levels increase in cycle 2 and 3.

This baseline level for awareness will be constructed by interpreting the outcomes from user story workshops in month 14 (Chap. 3). For example, for perception, we infer from the user stories that the users only perceive the threat when an employee comes to them with an issue (not immediate), they have to research what caused the issue (no immediate diagnosis) and what kind of threat is causing the issue. This will give no points for immediate perception of a threat. For comprehension, some baseline understanding may be assumed, once the threat has been diagnosed, but not all details will be known. For projection, the measures will be adequate, but probably too rigorous, even at the level of not allowing certain applications to be used. For decision, some interaction and communication policies (National Institute of Standards and Technology, 2018) have been implemented, but not always, and not in every case. Similar conjectures apply to sharing and self-healing. We will look into this more precisely when we set the baseline for the two municipalities (Sect. 5.5.4).

5.4.1.4   Level 4: Organisation

To what extent is the information that system provides also exploited elsewhere in the organisation, are the managers aware of this use, and what is its impact?

The requirements for organisational awareness link to the general requirement of the CS-AWARE system: to provide cybersecurity awareness (S1), as well as to the system requirements set for compliance (S10) and Integratability (S11). We distinguish awareness of impact on the municipality network, on services for citizens, and on organisational culture (Anttila & Knowledgist, 2006):

- O1 (Impact on the system): one aspect of organisational awareness is related to increased security of municipality data, which is their greatest asset. Also, we ask about compliance with local regulations, and about improved security and resilience in general. These KPIs are related to S10 and S11.
- O2 (Impact on citizens): this concerns the extent to which more effective delivery of services is made possible by using CS-AWARE. It concerns S1 and S11.
- O3 (Impact on culture): we inquire about the extent to which senior management is more aware of cybersecurity, and the impact on general organisational culture and security awareness. This involves requirements S1, S10 and S11.

We will administer Questionnaire 4 to the managers within the organisation. These will be managers of the system department, but also managers from the departments involved in the pilot scenarios that have been selected for CS-AWARE.

5.4.2   Qualitative Instruments and Procedures

In the previous section we addressed how requirements and subsequent KPI's are addressed in the questionnaires for each level. In this section we will describe other evaluation instruments and the qualitative ways adopted for some of the levels, within our hybrid evaluation approach.

Level 1, Technical: The main input comes from members of the deployment team commenting on technical details, during team meetings, or through email.

Level 2, Usability: The purpose of usability testing was to get hands-on information on how users handle the CS-AWARE interface in threat situations. Tests were run in a test environment which presented a number of cybersecurity incidents that were typical instances of the use cases. Usability testing involves representative system users who are subjected to a thinking aloud procedure, inspired by a cognitive walkthrough approach (Mahatody et al., 2010). After a brief introduction, the participants are invited to perform a number of simulations defined according to the use cases. Participants are asked to express aloud their thoughts, feelings and opinions on any aspect during their activity. Ideally, users perform the tasks without more explanation other than the brief introduction; at the end of each task, they can provide their main comments. Once the participants have completed the tasks, they are invited to fill in questionnaires 2 and 3. The verbalisations of the participants and their corresponding usage of the interface were all recorded and transcribed for analysis, with their permission.

Level 3, Awareness, qualitative analysis: early development of awareness, i.e., during cycle 1, was interpreted by comparing the base-level (set by the story workshop, see Chap. 3) with the deployment scenarios that were constructed by the deployment teams of Larissa and Rome (discussed in Sect. 5.3, constructed in month 26 of the project). Questionnaire 3 was administered to participants immediately after the usability test in cycle 2 and 3.

Level 4, Organisational: No other qualitative instruments were used for level 4.

Questionnaire 4 was sent to the managers of the two LPAs. This was done twice; at the end of cycle 2 and at the end of cycle 3.

Differences between cycles: We should note that the three cycles of deployment involved different deployment activities, and, consequently, they were not identical in terms of evaluation. For cycle 1, we performed technical and qualitative usability evaluation, and compared user awareness between what was revealed in the user stories (month 14, baseline) and the deployment scenario (month 26). This will be reported in the outcomes section for cycle 1 below.

For cycles 2 and 3, all levels were formally tested by administering questionnaires, and qualitatively by organising two more rounds of usability testing.

Table 5.6 provides a summary of the methods and instruments used for evaluation in this project.

5.5   Deployment and Evaluation Outcomes in Cycle 1

The first deployment cycle took place between months 26 and 28 of the project (Oct-Dec 2019). The goal of this cycle was to implement and test the system in the LPA contexts, to plan feedback procedures with the users, and to collect initial measures for evaluation.

During cycle1, implementation of the system in the local network was still ongoing, therefore we focused on the first three evaluation levels: technical, usability, and awareness. Awareness was approached through the interpretation of the user stories (baseline setting) and the interpretation of user objectives revealed in the deployment scenarios (qualitative analysis of awareness). Table 5.7 shows the activities undertaken during deployment cycle 1. We first discuss our deployment activities, and then report the outcomes for evaluation.

5.5.1   Deployment Activity in Cycle 1

Deployment Preparation: Deployment at the CS-AWARE project level started in May 2019. We formed a Pilot Working Group, which had weekly meetings. A template for a deployment scenario (Sect. 5.3) was proposed and discussed. We then specified the evaluation plan, which included the negotiated timing of all validation activities. We worked on interface usability testing by engaging a number of colleagues from the CS-AWARE project in a number of small tests.

At the beginning of September 2019, a workshop was organised to discuss expectations from the CS-AWARE partners about deployment, use, evaluation, and

Table 5.7 Activities for deployment and evaluation during cycle 1 (October–December 2019)

CYCLE 1	M26	M27	M28
Deployment			
Technical	System development		
Users	Deployment team and scenario (L)	Deployment team and scenario (R)	Deployment team meetings
Evaluation			
Level 1: Technical			System test
Level 2: Usability			Testing (L-R)
Level 3: Awareness		Qualitative baseline setting / analysis of awareness (L-R)	

impact. We discussed the feedback we would be collecting from the users. Furthermore, we addressed issues at the organisational level, both from the user point of view (how would experiences with the CS-AWARE solution be disseminated inside the municipality?) and from the project point of view (can the system generate information that can be exploited by the organisation? How can this be disseminated and exploited?). We also discussed the possibility of the changing role of the system administrator, when awareness is increased (e.g., more or different responsibilities), and, accordingly, more accurate and effective services that can be provided. All these points of discussion fed into finalising the deployment scenario template, and also in our understanding of the current level of awareness of the future users of the CS-AWARE system in Rome and Larissa who were present at this meeting and participated at the workshop.

Deployment scenarios started in October 2019, with two workshops together with the users from both municipalities. The main outcomes of these workshops can be found in Tables 5.2 and 5.3. The two workshops provided excellent insight into the objectives of the users at the two LPA pilot sites.

Deployment activities with users: After the start of the deployment scenario, we had three meetings with the deployment teams in Larissa and Rome. We further discussed the deployment scenario, in addition to details of the CS-AWARE interface. In addition, there were the discussions about technical implementation, taking place on a weekly basis, and through frequent email exchanges. We organised formal usability tests for evaluation in December in which three users (all members of the deployment teams) participated at each location.

It should be stressed that all documents and comments were discussed with Larissa and Rome extensively and are published with their full consent.

5.5.2   Technical Validation during Cycle 1

The technical validation focuses on testing the functional requirements for each individual CS-AWARE component (see Chap. 4: system and dependency analysis – GraphingWiki, data collection and storage, data pre-processing, data analysis and pattern recognition, multi-language support, visualization, information sharing and self-healing), as well based on the functional requirements defined for each component.

The test results showed that all basic technical functional requirements defined by CS-AWARE have been fulfilled. Work continued on refining the technical basis based on input derived from CS-AWARE piloting.

5.5.3   Usability Testing in Cycle 1

Feedback on usability was a very productive line of investigation and resulted in many important improvements of the CS-AWARE system.

Developments on the CS-AWARE console (i.e. the interface for users to the CS-AWARE system) were (up until cycle 3) subject to an ongoing sequence of revision and feedback, based on comments, from users, but also from the technical developers and other project team members. The results that we report here as outcomes of cycle 1, therefore pertain to a version of the interface that no longer exists, although the structure of the CS-AWARE interface is still the same (Fig. 5.3).

Table 5.8 shows the main features of the usability tests in both Rome and Larissa. As can be seen, there were three participants in the usability testing at each LPA. In Larissa, participants were tested individually, in Rome, the participants were tested with the active presence of their colleagues and a translator from CS-AWARE. We have explained in the methodology Sect. 5.4 that we approached threat detection, comprehension and resolution as a decision-making process, comprising 4 phases. A schematic overview of the phases and the appropriate screens of the console can be seen in Fig. 5.4. Here, we will discuss the main changes made based on user feedback, for each of the four phases of the decision-making process (see Sect. 5.4.1.3).

Phase 1, perception: This refers to the opening screen showing a dartboard shape with current threat types, in different colours, reflecting their estimated urgency. These colours were simplified for the sake of clarity (Fig. 5.3). The opening screen also displays a table with the list of threats, with additional characteristics, derived from trusted information sources. This was much appreciated by the users, and some of the details were improved, for clarity of description and easier identification of threats in the list.

Phase 2, Comprehension: The threat information window (see Fig. 5.4) is revealed when the user clicks on a threat in the opening screen. It shows more information about that particular threat. Graphics were improved, and a comprehensive list of technical details of the threat for system administrators was added.

Phase 3, Projection: In the network visualisation, users can focus on the components of the system network, the nodes that are affected, and information flows for

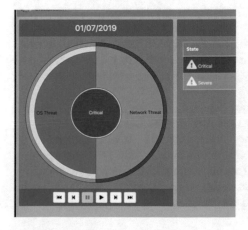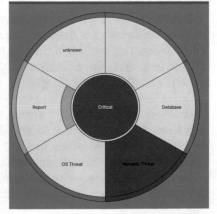

Fig. 5.3 Simplified interface for threat detection. Initial version on the left, revised version on the right

Table 5.8 Main features of usability tests in cycle 1

Summary	Usability study, Rome & Larissa, cycle 1
Use cases	(1) A vulnerability use case; the network has been infected. Tasks: Locate the source, find the infected parts, and undertake appropriate action, including deciding on self-healing. (2) A general security warning, such as a DDoS attack. (3) A malicious IP address has been detected. (4) An attack against LPA data. For example: Unusual behaviour in security log, network traffic, database, at different moments in time.
Components	All components of the system are involved
Participants (Rome)	Two system administrators (SUET), one manager (head of the data Center of Roma Capitale)
Participants (Larissa)	Three system administrators, specialised in applications or in database services
Method	Cognitive walkthrough
Outcomes	Improvements (see text); better understanding of impact of expertise (role) and organisational complexity

business processes involved (Fig. 5.4). More details of the system nodes were provided at the request of the users. We added the possibility of threat filtering (by group, person, location) and the possibility of focusing on a single flow of information (e.g. for finance). Zooming was improved, search functions were added, and the shapes of the nodes in the visualisation more consistency matched their function in the network.

Phase 4, Decision-making: Decision-making in CS-AWARE entails communicating threat information to the stakeholders who are involved in repair. This can be done in the threat information window. The most important addition, at the request of the users, was a ticketing system for assigning roles, which is especially useful for a large municipality. This was a consequence of the differences in user roles and expertise of participants that we observed in the usability tests. Moreover, the interface for information sharing (not shown here) was enhanced allowing more detailed sharing options.

Differences between municipalities: The thinking aloud approach revealed important differences in the ways that the individual participants handled cybersecurity issues. This confirmed what we already understood from the workshops, and allowed us to make these differences in approach explicit, in particular with respect to the phase of decision-making.

In Larissa, we worked with experienced system administrators, who had sufficient autonomy to decide if a threat was resolved or not. During the full process, they were already used to consulting their colleagues and manager, depending on the kind of threat. We noted that system administrators differ in their expertise: specialisation in database administration, or in handling issues with particular applications. This difference in expertise was reflected in decision-making, where the appropriate expert was always consulted before a decision.

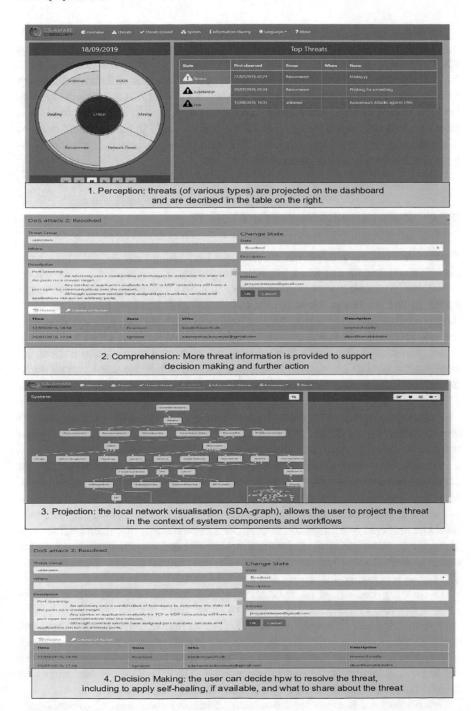

Fig. 5.4   Scenario (4 phases) for cybersecurity decision-making linked to features of the interface

In Rome, a similar difference in expertise appeared between database management and management of specific services (often outsourced), but there was an additional layer of decision-making. A new threat was always handed over to the expert responsible for the affected part of the system network. This is a crucial step in a context where permissions are distributed and some of the services are outsourced to commercial providers. Decision making therefore always involved a manager with the responsibility for handing over the issue to the dedicated experts. The manager was very knowledgeable, did not resolve the issues, but was responsible for deciding if a threat was resolved or not.

5.5.4   Awareness Outcomes for Cycle 1

Awareness baseline setting: For baseline understanding how the two municipal organisations handle and think about cybersecurity, we organised a story-telling workshop. We think stories are a good way to capture personal views and experiences. In our workshop, we added the collaborative dimension, because we think that a collaborative story (a story written together) can be taken as joint understanding and therefore constitutes a coherent integration of personal views and experiences. The stories therefore can be interpreted as providing a picture of how the organisation (and its employees) look at cybersecurity, in other words, on their initial level of awareness. The stories and their interpretations are reported in detail in Chap. 3.

Here, we recapitulate the main characteristics of the two LPA contexts, as they were revealed by the stories. In Larissa, there is one system administration department for the municipality (including smaller municipalities around Larissa), where all cybersecurity issues are handled. System administrators in Larissa have different expertise, are aware of each other's competences, and regularly share information. Managers from other departments seem or need to be less involved. In Rome, there are many departments responsible for many services, making for a complex structure. Departments handling data, or particular applications, or citizen internet services, do not necessarily share information on cybersecurity. It is unclear to what extent communication between the various departments takes place. For awareness, this means it is highly distributed, and often unshared.

Developing knowledge on cybersecurity is a local affair, and this is fine in Larissa, but less so in Rome. Application of safety rules is monitored in Larissa by the system administration, but it is less clear if other department managers share this attitude. Monitoring is not by policy, but is a collateral of users reporting issues. In Rome, if and how safety regulations are monitored, is highly dependent on the department managers, and is no consistent policy. The role of system administrators is to help users, sometimes even to help them to avoid regulations.

From this brief analysis we learnt that increasing awareness of cybersecurity issues and their resolution within system administration involves: (a) having better access to the knowledge of threats (from various sources); (b) having an interest not only in threat resolution, but also on the implications for the system network of the organisation; (c) having an interest in safety of the department and the organisation; and (d) having an interest in communicating and sharing experiences with other municipalities, CERTs and NIS authorities.

The rules and norms that the IT-Department maintains underlying the regulations for monitoring safety behaviour, and providing services to the users, are different in the two municipalities. In Larissa, we see that the system administration expects citizens (internal users) to stick to the rules and regulations. Conversely, the citizens expect an immaculate and timely resolution of their cyber-issues. This means there is always a tension: trust may grow but also decrease. In Rome, there seems to be a general norm that the system as a whole is too complex, and we should accept imperfection to the extent that not all policy regulations are created in the interest of maximal cybersecurity. It looks like the various IT-departments are still able to provide quick services to users with problems. These services are provided on an individual basis, for a user who detects an issue, there is no follow-up, and there may be no spread of information to other users in case of a security issue. There is a lack of awareness of threats, especially in terms of understanding the threat and its impact on the system network. In Rome, this lack of awareness is related to the distributed knowledge about the complex system network: there is no single individual who completely oversees the whole network.

Concerning sharing of information within the organisation, in case of Rome, there is the issue of information sharing and knowledge management: in Rome, knowledge is highly distributed and less shared, system users seem to have no concern for each other's cybersecurity issues, and the culture favours general regulations but individual solutions. For awareness, this creates a problem: it does not evolve nor spread. Also, in Larissa, while employees within the system department are aware of the cyberincidents that have occurred, this knowledge remains within the department, and seems of less concern for others, although the other employees that we met seemed to have some interest.

Although the situation in Rome reflects a more complex organisation for handling cybersecurity, underlining the importance for managing and sharing information, the main characteristics of awareness for Rome and Larissa are (perhaps surprisingly) similar. In general, we understand that cybersecurity awareness always is a collaborative effort, it involves the responsibility for all users to share and act together. When knowledge is distributed, sophisticated management and collaboration are essential (Andriessen & Pardijs, 2021).

5.5.4.1   Awareness Requirements, Baseline Level

For interpreting the baseline level of awareness, which is the state of cybersecurity awareness at the beginning of the project, we combine the scenario of cybersecurity decision-making with inferences from the user stories. In our discussion, we focus on system administrators (level 3 of evaluation), but also on relevant aspects of the organisation (level 4).

A1, Perception (The 'A' in 'A1' signifies 'awareness'): Threat perception at baseline is not immediate, as it usually depends on the system administrator (the user) being informed of an incident (by a service user, or a system alert). Then, the user has to identify the threat, discover its possible impact, such as inferring when the issue probably appeared, in what part of the municipality network, or with which employees. Our understanding is that the focus in both municipalities is on threat identification, and less on the other processes (comprehension, projection, decision-making). Base-level perception is not immediate, identification can be time-consuming, focuses on individual service users, and is local (part of the system), not (whole) system oriented.

A2, Comprehension: For further understanding of a threat, if it is not already well known, a system administrator has to access the same resources as for identification. Searching for and reading relevant information may take time, depending on the experience of the administrator, and often does not happen at all. The administrator may have to research log-files of the system, and maybe of particular services, to figure out what exactly is the problem. System administrators may do this, if they are responsible for the network or a part of it. If their role is to manage the database of a particular service, they may not be interested in such log files. In that case, communication with another specialist is needed, for exchange of information about the threat to be collected and understood.

For awareness this means that for a system administrator resolving a threat, full comprehension is not always necessary, and sharing information is only needed sometimes. Therefore, comprehension is limited, both at the individual and at the organisational levels.

A3, Projection: If the system administrator wants to understand the risks imposed by a threat to a system node, as well as to the other nodes, and which services will be in danger, a thorough understanding of the network as well as of the information flow in which the affected nodes are involved is a requirement. As awareness often is distributed, these aspects of resolution are a problem for users, especially in larger organisations. Users may be very knowledgeable about their own service, but may not always oversee the possible implications for the rest of the network.

For awareness, this means that projection is local, and users may lack the knowledge to research their network. In Larissa, communicating with colleagues often solved the problem. In Rome, such communication may be more complicated, and require the intervention of management.

A4: Decision: The final phase is called decision-making, and this involves several different actions. In the baseline situation, it usually means resolving the threat, for example by applying a patch or update, or blocking a user or part of the network. In a complex organisation, it may mean referring resolution to the expert responsible for that part or service in the network. In all of these situations, experts, colleagues, including managers, may be informed or consulted. This requires sufficient comprehension. In addition, somebody may need to ascertain that a threat is actually resolved. Most awareness in the base-level situation depends on communication between different stakeholders. In other words, awareness requires the collaborative attitude of all making contributions, based on a desire for sharing, and careful consideration of all contributions.

5.5.4.2 Qualitative Interpretation of Awareness

The deployment scenarios started at the beginning of the third year of the project, hence 1 year after the story workshop. Compared to the user stories collected at the start of year 2, we can see (Tables 5.2 and 5.3) increased user awareness of cybersecurity, reflected as (1) Perception: expectations for timely and accurate detection of cybersecurity threats; (2) Comprehension: awareness of the importance of good quality information and good quality reporting; (3) Projection: interest in their own system network, its assets and vulnerabilities, and the additional possibility of reflection on past problems and solutions; (4) Decision: the possibility of mitigation reports and user responsibility in making decisions; (5) Interest in impact on the organisation and communication with internal and external service users.

Nevertheless, we also see aspects where awareness gains are quite possible: (1) The focus still is on detection rather than on comprehension; (2) Especially in Larissa, trust in the CS-AWARE system still needs to evolve; (3) Users did not discuss sharing (with competent authorities) and self-healing; (4) Efficiency rather than awareness is the main focus; (5) No additional policies for monitoring and sharing (within the organisation) have been addressed.

Also, we note the following differences between Rome and Larissa, some of which may be related to differences in the complexity of their organisations: (1) Department-oriented objectives (L) versus organisation oriented objectives (R); (2) Orientation towards greater efficiency and trust (L) versus collaboration between managers and between the different departments (R); (3) Service users are interested (L) versus not interested (R) in security; (4) System administrators stress the importance of being alerted (R) versus being involved in active learning and discussion (L), and, finally, (5) higher involvement and expectations by either managers (R) or by system administrators (L). One comment about (2) orientation is, that this orientation signifies policies that are not yet realised in the department or organisation.

Awareness development, cycle 1: The two Tables 5.9 and 5.10 summarise the differences noted between the awareness of cybersecurity at the beginning of the project (month 14) and 1 year later (month 26), for system administrators in Rome and Larissa. It should be noted that when system administrators or managers discuss some issue, identified by the CS-AWARE system, in a deployment team, this does not imply that this awareness is always applied when resolving a threat. For example, the most efficient use of the CS-AWARE system is possible by simply

Table 5.9 Awareness baseline, and at the start of deployment (Rome, system administrators)

Rome (sysadmin)	Month 14 (story workshop)	Month 26 (Cycle1)
Threat perception	Not immediate, local, individual	Real time alerts, efficiency
Threat comprehension	Distributed expertise, distributed rights & roles	Efficiency
System projection	Service (not network) oriented, requires management	Service (not network) oriented
Decision-making	Distributed	Trouble ticket, more effective relations
Need for sharing	Manager distributes tasks	Better communication with internal users
Self-healing	=	=
General	Individual help, no follow-up	Reflection

Table 5.10 Awareness baseline, and at the start of deployment (Larissa, system administrators)

Larissa (sysadmin)	Month 14 (story workshop)	Month 26 (Cycle 1)
Threat perception	Not immediate, local, time consuming	Immediate, no false alarms
Threat comprehension	Only if needed, collaboration	Up to date, if trusted then collaboration
System projection	Collaborative	Proactive, reports, collaborative
Decision-making	Collaborative	Informed, self-paced
Need for sharing	Internal	Internal
Self-healing	=	=
General	Trust between sysadmin and user is an issue	Learning

confirming current practice: nothing changes, all remains the same. In that case, although detection of a threat is improved with CS- AWARE (automatic notification and diagnosis of threats), the user can simply follow the suggestions provided by the system until the threat is resolved, with only minimal reflection.

In terms of increased awareness of cybersecurity threats, the notion of (better) threat comprehension is therefore very important. In both LPAs, initially (in month 14) full comprehension of a threat was not a main goal, because the department was expected to handle threats as efficiently as possible. In month 26 users in Rome indicate they see the potential of CS-AWARE for (increased) efficiency. Greater efficiency in Rome is a general organisational desire, which may include cybersecurity. The managers in Rome were clearly interested in reflection on how their organisation handled cybersecurity. In Larissa the users indicated they were keen to learn from using CS-AWARE, but only once they had developed sufficient trust in the information that the system provides.

Concerning projection, although none of the users in Rome referred to the need for understanding the system network, all agreed that more effective communication was needed. In Larissa, the users expressed the desire to have overviews of what system components were threatened, and in collecting evidence about the weaker components in their network.

As for decision-making, the users in Larissa indicated they expected decision-making to be better informed with the CS-AWARE system, and this would also support internal collaboration. In Rome, high expectations were formulated for better communication with service providers, and more effective management of the various users involved in threat resolution processes.

The need for sharing only was mentioned as an LPA-internal possibility, not with external communities. Self-healing was not mentioned at all.

In our qualitative approach we interpreted the meaning of user stories and deployment scenarios for awareness. Clearly, we are interpreting and evaluating a developmental trajectory for awareness, not a fixed state. Our conclusion is that, for awareness, we could already see clear developments of awareness, on both sites, between month 14 and month 26. For Rome, this means increased awareness of the importance of CS-AWARE for threat perception and comprehension, and especially, for decision-making.

5.5.5   Conclusions of Evaluation of Cycle 1

Achievements in cycle 1: Table 5.11 summarises the outcomes for cycle 1. The CS-AWARE system implementation was finalised and deployment in the LPA system networks started at the end of the cycle. The deployment teams were established, which meant we had installed a group practice for feedback and testing. The deployment scenarios made clear what our participants expected and the state they were in concerning cybersecurity awareness. Usability testing was a very productive source of information, not only for the system, but also for understanding its use by the participants. We discovered many differences in use, linked to the expertise and role of the participants in the organisation. We observed a considerable increase in awareness of cybersecurity during months 14 and 26, which could be linked to participation in the design of the CS-AWARE system, and the workshops (Chap. 2) in particular.

Lessons learnt from cycle 1: In this project, we are designing a bespoke solution for cybersecurity awareness for two LPAs. The assets of this solution are clear: the system works in the local context, and for the local users. The limitations are clear as well: the approach requires user investment, and generalisation from two pilots to other pilots seems difficult. Nevertheless, our qualitative approach to deployment has given us information on the development of awareness of cybersecurity at the two municipalities.

Table 5.11 Summary of outcomes of cycle 1 for the two LPAs

Deployment	Rome	Larissa
Technical achievements	Data center requirements gathered; Enterprise docker set-up; IAM integration	Cloud requirements gathered; full deployment on AWS
Deployment team	13 participants	6 participants
Objectives	Efficient detection, better collaboration	Effective detection
Foreseen impact	Improved communication with service providers	Learning, reputation
Desired	Alerting and reporting	Reporting
Evaluation	Rome	Larissa
(1) Technical		
Method	Functional component and integration test set	
Outcomes	Full set of functional tests (component and integration) passed	
(2) Usability		
Participants	3 (1 data center manager and 2 sysadmin)	3 (system admin)
Method	Cognitive walkthrough	Cognitive walkthrough
Outcomes	List of recommendations	List of recommendations
Processes	Distributed	Collaborative
(3) Awareness		
Method	Story telling	Story telling
Participants	15, 5 stories	13, 6 stories
Awareness baseline	Local, individual	Local, individual
Participants	13	6
Method	Co-creation	Co-creation
More aware of	Threats, network, organisation	Threats, need for trust
Not more aware of	Sharing, self-healing	Sharing, self-healing

We propose here, on the basis of the user stories, the following definition of cybersecurity awareness that we will further exploit during cycles 2 and 3:

■Awareness of cybersecurity includes both knowledge and agency.
Knowledge pertains to:
   1. Cybersecurity threats
   2. The system networks
   3. The organisation and its users and
   4. The cybersecurity community.
Agency is about the ability and willingness to act:
   5. When a threat is imminent
   6. When there is no threat

We suppose that the categories (each to a certain degree) qualify user-awareness, its role during detection, but also when there is no threat to resolve. We do not test detailed knowledge of cybersecurity threats directly, but we will test, during action to resolve a threat (the usability test), and after that action (questionnaires 2 and 3) the extent to which the user specifies that a particular threat is understood, including the implications for the system network, and what to share with the cybersecurity community. In addition, we ask about (in questionnaire 4) the implications for the organisation: the system network, the cybersecurity culture, and the impact on citizens.

The CS-AWARE system provides to a user knowledge about threats, about the system network, and the components and service processes implied in the threat. This may lead to increased awareness of cybersecurity in the organisation. CS-AWARE supports the user agency when resolving a threat, and this may facilitate awareness and reflection also when there is no immediate threat. The CS-AWARE system asks users about what they want to share about a threat with cybersecurity communities. This requires a dedicated policy for sharing at the municipality.

So, as a general expectation, all aspects of awareness may improve, but not automatically. In that respect, we note the important concept of sharing that will be the focus for the next phase.

Implications for next phase: During the next phase, we will focus on the implementation and testing of information sharing of threats. Sharing such information to CRTs and other stakeholders does not have a high priority for our LPAs at the moment. Furthermore, we feel safe now to build on our understanding of cybersecurity awareness at the pilot sites to undertake a quantitative evaluation in the form of questionnaires.

5.6 Deployment and Validation Outcomes for Cycle 2

The goals of cycle 2 were achieving a full deployment of the CS-AWARE system at the user sites, collecting further user feedback, getting results on the questionnaires for all levels, and to perform further usability tests. Table 5.12 shows the activities in cycle 2.

5.6.1 Deployment Activities in Cycle 2

Based on the already existing (and tested) functionalities, work continued for preparation of a test of three main use cases in the deployments in Larissa and Rome. This consisted of work to instantiate the specific use cases, and refine the components involved in the cases along the way:

Table 5.12 Activities for deployment and validation during cycle 2

Timing	M29	M30	M31
Deployment			
Technical	System development		
Users	Project meeting	Review meeting	Deployment team meetings
Evaluation			
Level 1: Technical			Questionnaire 1
Level 2: Usability			Usability test Questionnaire 2
Level 3: Awareness			Questionnaire 3
Level 4: Organisation			Questionnaire 4

(1) *Instantiation of Rome/Larissa specific monitoring patterns*: The involved components in this use case are system and dependency analysis, data collection, data pre-processing, data analysis and visualization. The purpose of this use case was to instantiate the monitoring patterns derived during the third system and dependency analysis workshops in Larissa and Rome (Sect. 5.2), to be able to test realistic threat events in the context of the "suspicious behaviour monitoring" use case. The following work on the individual components was required to provide this use case:

- System and dependency analysis: It became clear that the asset and dependency graph format required, in addition to be able to model asset and dependency information, additions to be able to also model information relating to log files and monitoring patterns.
- Data collection and data pre-processing: The log files for Rome and Larissa were already collected and pre-processed in earlier phases. Small refinements and bug fixes to individual log file parameter processing have been implemented on a per-case basis.
- Data analysis: The Rome/Larissa monitoring patterns have been implemented by the data analysis component.
- Data visualization: Minor refinements and bug fixes have been implemented in the process of visualizing the Rome/Larissa specific patterns

(2) *Information sharing*: Based on the information provided by the Rome/Larissa specific monitoring patterns, information sharing functionality has been refined in cycle 2. The involved components have been information sharing and visualization. The main effort has been to rework the communication protocol between the information sharing and visualization components, based on the experiences with the data from the real Rome/Larissa use case patterns. Figures 5.5 and 5.6 show examples of the main screens for information sharing.

(3) *Social media monitoring* (general security warnings use case): A concrete implementation of the general security warnings use case was provided in cycle 2. It was decided to allow the user to monitor social media messages for specific keywords on a per asset basis. If a keyword is detected, it is displayed as a "social media

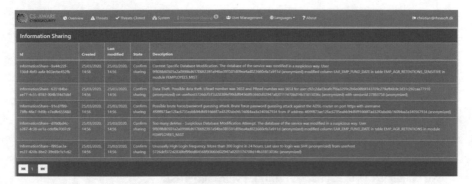

Fig. 5.5 Information Sharing overview

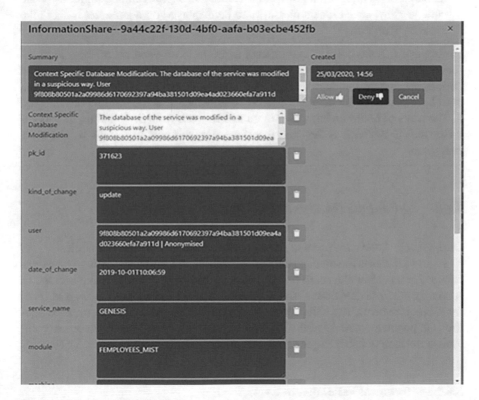

Fig. 5.6 Information sharing details

report" with default "low" threat level in the CS-AWARE overview, and visualized at the asset for which the keyword was specified. The involved components are system and dependency analysis, data pre-processing, data analysis and data visualization:

- System and dependency analysis: Keywords are modelled in the asset and dependency graph in the already foreseen "categories" parameter. No changes were required.
- Data pre-processing: A specific pre-processing task was added that allows data analysis to trigger a keyword-based search in a specified set collected information form information sources (like e.g. social media), based on the keywords defined in the asset and dependency graph.
- Data analysis: The ability to compile social media reports from messages that were retrieved from keyword-based search, in line with the previous protocol (where threat type is "report" and priority is "low" by default), was implemented. No changes to the exchange protocol were required.
- Data visualization: Representation of social media threats is in line with the visualization concept, no changes were required—except for minor additions to the code to allow social media messages to be displayed on a per-asset basis in the systems overview.

During cycle 2, we collected comments from users on versions of the additional functionalities mentioned above. It should be noted that cycle 2 was a turbulent period, because of the start of the covid pandemic, which halted communication with Larissa during the month of March 2020, which was the third month of cycle 2. The users of both municipalities often had to work from their homes, and monitoring the CS-AWARE system was therefore more difficult, especially for the users in Larissa.

5.6.2   Validation Outcomes in Cycle 2

During cycle 2, we administered the 4 questionnaires in both municipalities.

Figure 5.7 shows the outcomes for each requirement, collapsed over all questionnaires. For example, the score for S2 (allow information sharing) is the average of scores for S2 from questionnaires Q1 (system) and Q3 (awareness). As can be seen, for all requirements, the KPIs are above threshold (.6), so the scores seem to confirm the positive outcome of the CS-AWARE system deployment. Below, we will look at the results for each questionnaire in more detail.

5.6.2.1   Technical Validation in Cycle 2

Questionnaire 1 was administered to users in Larissa and Rome. Table 5.13 shows the outcomes for each municipality, separated for type of user and the requirements that were tested.

The following observations seem relevant, where we take a difference in scores of more than 20% as noteworthy:

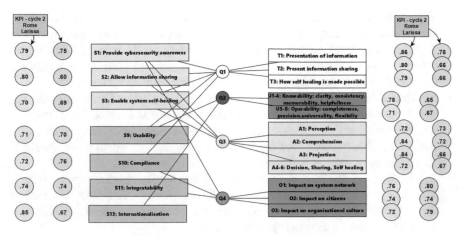

Fig. 5.7 Overview of questionnaire scores in cycle 2. Scores on the left are averages of scores on the right, for each requirement

Table 5.13 Outcomes for Questionnaire 1 (cycle 2), for each requirement, in proportion of users (0–1). Exp means experience of the user (1–5)

Q1-CYCLE 2	Sysadmin	Managers	Service users	Mean
Rome				
N	4	10	6	20
Exp	2	1.5	1.8	1.7
S1	0.74	0.82	0.75	0.78
S2	0.57	0.72	0.61	0.66
S3	0.6	0.72	0.59	0.66
S9	0.6	0.78	0.73	0.73
S10	0.6	0.70	0.67	0.72
Mean	0.64	0.76	0.67	0.71
Larissa				
N	3	0	3	6
Exp	5		2	4.67
S1	0.82		0.9	0.86
S2	0.56		0.91	0.69
S3	0.8		0.87	0.84
S9	0.8		0.93	0.87
S10	0.8		0.87	0.83
Mean	0.74		0.9	0.81

- The KPIs set for all requirements overall are 60%, which means that, at this stage, the key performance indicators confirm that the requirements for S1 (provide cybersecurity awareness), S2 (allow information sharing), S3 (provide self-healing), S9 (usability), and S10 (compliance) are met.

- For Rome, there is a relatively high representation of management in the participants who filled out questionnaire 1. These managers generally assigned somewhat higher scores than their system administrators (on average 15% higher) on all requirements. At the same time, both managers and system administrators indicate they had not spent much time in working with the system yet.
- For Rome, the ratings by system administrators are, in spite of their lack of experience with the system, lower than the ratings by their managers, for all requirements (see Table 5.13), with the exception of S1 (provide cybersecurity awareness). Their comments on features of sharing information, self-healing, and usability in general will be collected in the usability test.
- For Larissa, we see highly experienced system administrators, who give high scores to each of the requirements, with the exception of S2 (allow information sharing). We take the combination of high scores and high experience with CS-AWARE as a very encouraging result
- Requirement S2 receives low ratings (below 60%) from system administrators from both municipalities. We will explain this low score in the usability section.
- Service users at both LPAs evaluate the system quite positively, but they do not have hands-on experience, these scores should be taken as the result of them participating in workshops and witnessing a demo. For them, CS-AWARE 'looks good'.

The outcomes of Q1 are positive overall, although there are differences between managers (more positive) and system administrators in Rome, the system administrators in Larissa have more experience and are more positive, and it is worthwhile to inspect the process of sharing information (at the interface) more closely.

5.6.2.2 Usability Testing and Questionnaires in Cycle 2

In Larissa, due to the situation of citizens/workers being locked at home, the users worked in their own time on testing the interface. They provided a set of written comments, that were further discussed with them in a deployment team session.

In Rome, we organised a number of usability tests, involving four different users. With the first three users, we noted that it was impossible for a system administrator, for example responsible for maintaining a service database, to make decisions about how to resolve a threat, because that decision was the prerogative of a manager with overview over the municipal network. We therefore organised an additional session, in which the manager and the system administrator together resolved a use case.

We then analysed in more detail a single use case, with respect to the three new features (discussed in Sect. 5.6.1): a specific monitoring pattern triggered a critical warning, and in addition to resolving the threat, users were requested to comment on the detailed pattern information, the options for sharing, and, as an additional feature, the output from key word search on social media that the system had performed on user provided keywords.

The fictional threat was 'injected' into the deployed CS-AWARE system. Participants in Rome were recorded when they were thinking out loud in their attempts to resolve the threat. All components of the system are involved, we had specific interest for LPA specific monitoring patterns, details of information sharing, and keyword search. Table 5.14 summarises the usability tests in cycle 2.

In general, we noted that the sharing of information about the threat posed some problems for certain system administrators. Firstly, they were unaware with whom the information had to be shared. This was an unsettled issue indeed, and we suggested that the information that was marked as shared would end up in a database that could be accessed by people, internal and external, still to be designated, and with specific permission. Secondly, the information that was shared, was highly technical, to be understood by system administrators with the relevant knowledge. However, the decision to share or not was a management decision, it even required formal policies for sharing. Neither municipality had such policies installed (yet). Thirdly, all users were bothered by the (GDPR requirement, see Chap. 1) anonymity of the IP-Address, which, they thought, would be the most critical information to share.

To further understand if the information provided by the system for sharing was useful, in Rome we undertook another treat simulation with the CS-AWARE database manager and one system administrator together. This session revealed the central position of the manager, including his prerogative for decision-making. The system administrator (responsible for a particular application, for example) was assigned particular tasks by the manager and had to get back to him as soon as the task was accomplished. The manager decided if enough was done to resolve the threat, after communication with the administrators for all network components that could be affected by the threat. From the system, this required the extended possibility to send multiple messages to several administrators. It should be noted that the current medium for handling communication about threat resolution in Rome is a ticketing platform. Implementing CS-AWARE would require either integrating this platform in the communication process, or replacing its functionalities. The role of the manager is crucial for the success of the process.

Table 5.14 Main features of usability tests in cycle 2

	Usability study, Rome & Larissa, CYCLE 2
Use cases	A general security warning: a DDoS attack
Components	All components of the system are involved, we had specific interest for LPA specific monitoring patterns, details of information sharing, and keyword search.
Participants (Rome)	2 system administrators (database service), 1 manager (head of the data Center of Roma Capitale), 1 system administrator
Participants (Larissa)	3 system administrators, specialised in applications or in database services
Method	Cognitive walkthrough
Outcomes	A list of improvements

The users in Rome filled in questionnaire 2 immediately after the usability test, and the users in Larissa did the same during the last week of April, after individual testing of the interface. Table 5.15 displays the outcomes. The following comments can be made:

- Again, there is a difference in experience between the participants in Rome (low experience) and those in Larissa (high experience). These differences reflect the difference in complexity of the organisations, rather than underlying motivations
- Again, all scores a well above threshold (.6), except the score for Helpfulness in Rome, and the score for Flexibility in both municipalities. The mean score overall can be taken as evidence that the users evaluate the usability of the system as above average.
- The low score for Helpfulness in Rome means that the users indicate they needed more help than the system was providing. Our observations during usability support the explanation that system administrators need support for: (a) deciding that a threat is resolved (because more components could be involved); (2) interpreting information for sharing (because this requires a management policy); and (3) better understanding of possible actions in the system visualisation screen, and the operation of the keyword search, that was linked to particular system components.
- The Flexibility question involved the user's estimate that CS-AWARE was to be used by experienced users only. An affirmative answer to this question was scored negative on the scale for flexibility. A flexible system is a system that can be used by all types of users, experienced and less experienced. The low score here reflects the users' opinion that effective use of the system requires certain expertise, which is not necessary a negative outcome.

Table 5.15 Outcomes for Questionnaire 2 (cycle 2), for each requirement, in KPI (0–1). Exp means experience of the user (1–5)

Usability, cycle 2	Rome	Larissa
N	3	4
Exp	2	5
S9: Knowability		
Clarity	0.67	0.85
Consistency	0.76	0.85
Memorability	0.69	0.72
Helpfulness	0.5	0.70
S9: Operability		
Completeness	0.78	0.73
Precision	0.78	0.72
Internationalisation (S13)	0.67	0.85
Flexibility	0.47	0.55
Mean	0.66	0.75

We feel safe to conclude that, in spite of (or because of) the additions to the system interface requiring additional reflection and discussion, the score for Usability is above average. The scores on questionnaire 2 for usability are not very different from those of questionnaire 1, where usability was dealt with by one general question. The average scores for that question were 0.6 for Rome and 0.8 for Larissa, for system administrators.

5.6.2.3 Awareness in Cycle 2

The awareness questionnaire 3 was administered to the same users as questionnaire 2, after the usability test (for Rome) or after sufficient experience with the system (in Larissa). This questionnaire does not inquire about the success or ease of certain activities during the usability simulation, but on user awareness of the different aspects of the phases of decision-making. Users are therefore asked about the aspects of this process that they considered and took into account, and the communicative actions that they undertook. Table 5.16 gives an overview of the scores on awareness of each of the phases.

The relatively low score for sharing in Rome can be explained by the fact that users in Rome indicated they did not succeed in sharing, due to reasons explained under usability.

Figures 5.8 and 5.9 allow comparison of awareness scores between baseline, cycle 1 and cycle 2. Note, that the first two scores (baseline and cycle 1) were interpretations from qualitative information. As expected, all users in Rome indicate more awareness about threat perception and comprehension, which are strong assets of the CS-AWARE system. We see the same tendency for Larissa. Projection, or the awareness of the threat in the context of the system network, which is facilitated by

Table 5.16 Outcomes for Questionnaire 3 (cycle 2), for each requirement, in ratings (0–1). Exp means experience of the user (1–5)

Awareness, cycle 2	Rome	Larissa
N	3	4
Exp	2	5
S1: Provide cybersecurity awareness		
Perception	0.73	0.72
Comprehension	0.72	0.84
Projection	0.67	0.85
Decision	0.76	0.76
S2: Sharing		
Sharing	0.54	0.80
S3: Self-healing		
Self-healing	0.72	0.63
Mean	0.65	0.74

visualisation in CS-AWARE, also increased. We interpret this as a realisation in Rome of the importance of this aspect. In Larissa, in a less complex system, awareness of the system is more obvious. There, we see a highly developed awareness overall, with the exception for self-healing. This aspect will receive more attention during cycle 3.

As a conclusion we can say there was a satisfactory increase of awareness scores during cycle 2, especially in Larissa. This increase can be attributed to the many discussions about the system, and the time spent on working with the system, especially in Larissa. We will see if this proves sustainable in cycle 3, where the same instruments will be used.

5.6.2.4   The Organization Level in Cycle 2

Questionnaire 4 was filled in by 12 managers from Rome and by 4 system administrators and their manager in Larissa. Table 5.15 shows the results.

We can see in Table 5.17 the overall tendency of the managers in Rome to respond to the questionnaires with very positive ratings, especially for questions about the organisational level. Their experience with the CS-AWARE system may be low, but it can be supposed that there are many internal discussions in Rome about CS-AWARE, at the organisational level. This may be less the case in Larissa, but we see that the expectations (and experience) of the system administrators in Larissa are high for this level.

As a conclusion for the organisational level of evaluation we can say that KPIs are met, which is an excellent outcome with promising perspective for the actual success of CS-AWARE in the municipal context.

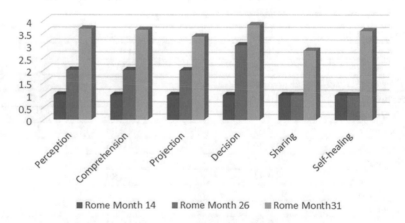

Fig. 5.8 Awareness test scores for Rome: before, cycle 1, and cycle 2

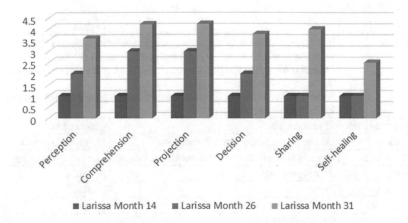

Fig. 5.9 Awareness test scores for Larissa: before, cycle 1, and cycle 2

Table 5.17 Ratings (0–1) 0f Questionnaire 4 (cycle 2), for each requirement. Exp means experience of the user (1–5)

Organisation, cycle 2	Rome	Larissa
N	12	5
Exp	1.8	4.4
S10: Compliance with internal regulations		
Perception	0.80	0.76
S10: Service delivery		
Comprehension	0.74	0.74
S1: Provide cybersecurity awareness (in organisation)		
Projection	0.79	0.72
Mean	0.78	0.74

5.6.3 Conclusions of Evaluation of Cycle 2

Achievements in cycle 2: Table 5.18 summarises the outcomes for cycle 2. Users from both teams indicated that the components of the deployment scenario, as established in November 2019, were still valid and no changes were necessary.

The specific monitoring patterns (for both Larissa and Rome) were instantiated, constructed from the input of the third round of SDA workshops. The visualisation and information sharing protocols were refined and better integrated. A new social media (general security warning) use case was instantiated at both locations. Through our usability testing we had good opportunities to test these additions and their impact on users.

The outcomes of the questionnaires overwhelmingly show that our intervention was well received at both locations. The most important difference between Rome and Larissa was the relatively high manager scores (and involvement) in Rome, and

Table 5.18 Summary of outcomes of cycle 2 for the two LPAs

Deployment	Rome	Larissa
Technical achievements	Instantiation of Roma and Larissa specific monitoring patterns finished Refinement of system and dependency graph protocol finished Internal information sharing communication protocol between information sharing and visualization component refined Social media monitoring (general security warning use case) instantiated for Rome and Larissa use cases	
Deployment team	13 participants	6 participants
Objectives	Efficient detection, better collaboration	Effective detection
Foreseen impact	Improved communication with service providers	Learning, reputation
Desired	Alerting and reporting	Reporting
Evaluation	Rome	Larissa
(1) Technical		
Method	Questionnaire 1	
Participants	4 sysadmin, 10 managers, 6 service users	3 sysadmin, 1 service user
Mean score	0.71	0.81
(2) Usability		
Method	Cognitive walkthrough	Cognitive walkthrough
Participants	3 (1 data center manager and 2 sysadmin)	3 (system admin)
Outcomes	Observations and implemented changes	Observations and implemented changes
Processes	Distributed	Collaborative
Method	Questionnaire 2	
Participants	2 sysadmin, 1 manager	4 sysadmin
Mean score	0.66	0.75
(3) Awareness		
Method	Questionnaire 3	
Participants	2 sysadmin, 1 manager	4 (sysadmin)
Mean score	0.65	0.74
(4) Organisation		
Method	Questionnaire 4	
Participants	12 (managers)	1 manager, 4 sysadmin
Mean score	0.78	0.74
(5) Business		
Method	Questionnaire 5	
Participants	None	None
Scores	NA	NA
Impression	No similar tool exists, there is a clear need	

the relatively high scores from system administrators in Larissa. In addition, in Larissa, the system administrators spent considerable time in appropriating the system, therefore, their feedback is considered especially valid.

The main goal of cycle 2 was to look at information sharing. This has technical implications (what to share about the threat) as well as organisational ones (who is allowed to share what to whom, and what is our policy?). In addition, there were privacy issues: what sensitive personal information is involved in sharing? We achieved to put this on the management agenda of the two municipalities, and agreed about a protocol, allowing the CS-AWARE tool to store 'shared' information in a database, with details to be selected by the users. This means that sharing works well at the technical level.

Lessons learnt in cycle 2: The corona-crisis had an impact on the availability of people for further feedback and testing. Methodologically, we saw that the questionnaires were returned reasonably well, in spite of these drawbacks. We learnt that the lifespan of a questionnaire probably is more than 3 months, and not bi-weekly, as we originally intended. Users indicated that they would not appreciate frequent consultation through questionnaires as nothing much would have changed within a cycle of 3 months. Whilst the outcomes of the questionnaires where encouraging, we learnt most from the usability testing, where we came to understand why our sharing of information protocol could pose problems to the users.

Implications for the next phase: As a general conclusion at the end of cycle 2, we have confidence that our approach is promising and has a good possibility for collecting further interest and potential new users, albeit at stages after the project.

5.7   Deployment and Validation in Cycle 3

The main goals of the third and final cycle of piloting (Table 5.19) were twofold. Firstly, to deploy the self-healing functionality at both municipalities, and second, to conduct final tests with the system.

Technical deployment: The technical developments in cycle 3 focused on the deployment of the last missing core feature of CS-AWARE – the self-healing feature—to the Rome and Larissa specific use case. This included a concrete instantiation of the self-healing policies for the Rome and Larissa specific monitoring patterns (instantiated in pilot cycle 2, as described in Sect. 5.6, and in Chap. 4, Sect. 4.2.1.7). The work in this cycle focused on instantiating the corresponding technical commands that would allow the policy to be implemented in specific cases.

Since it is seen as a bad practice to implement self-healing in the live systems of the Municipalities, a virtualised server environment was set up to act as the receiver and target of self-healing actions, instead of using the live systems of Rome and Larissa.

5.7.1   Validation of Technology in Cycle 3

Questionnaire 1 was again administered to users in Larissa and Rome. Table 5.20 shows the outcomes for each municipality, separated for type of user and the requirements that were tested. It should be noted that the response was lower than the first

Table 5.19 Activities for deployment and validation during cycle 3

Timing	M32	M33	M34
Deployment			
Technical	System development		
Users	Deployment team meetings		
Validation			
Level 1: Technical			Questionnaire 1
Level 2: Usability			Final usability test Questionnaire 2
Level 3: Awareness			Questionnaire 3
Level 4: Organisation			Questionnaire 4
Level 5: Business		Contacting new users and sending questionnaire 5	

Table 5.20 Outcomes for Questionnaire 1 in cycle 3 (cycle 2 between brackets), for each requirement, in proportion positive (0–1). Exp means experience of the user (1–5)

Questionnaire1 CYCLE3	Sysadmin	Managers	Service users	Mean
Rome				
N	1 (4)	4 (10)	5 (6)	10 (20)
Exp	2	1.5	1.8	1.7
S1	0.7 (0.74)	0.89 (0.82)	0.81 (0.75)	0.83 (0.78)
S2	0.73 (0.57)	0.77 (0.72)	0.72 (0.61)	0.74 (0.66)
S3	0.6 (0.6)	0.75 (0.72)	0.75 (0.59)	0.74 (0.66)
S9	0.8 (0.6)	0.85 (0.78)	0.76 (0.73)	0.8 (0.73)
S10	0.6 (0.6)	0.87 (0.70)	0.73 (0.67)	0.77 (0,72)
Mean	0.69 (0.64)	0.82 (0.76)	0.76 (0.67)	0.78 (0.71)
Larissa				
N	3 (3)	1 (0)	2 (3)	6 (6)
Exp	5		2	4.67
S1	0.88 (0.82)	0.80	0.93 (0.9)	0.88 (0.86)
S2	0.76 (0.56)	0.80	0.93 (0.91)	0.89 (0.69)
S3	0.83 (0.8)	0.80	1.0 (0.87)	0.87 (0.84)
S9	0.93 (0.8)	1.0	0.9 (0.93)	0.93 (0.87)
S10	0.87 (0.8)	1.0	0.9 (0.87)	0.9 (0.83)
Mean	0.84 (0.74)	0.84	0.94 (0.9)	0.89 (0.81)

time, due to questionnaire fatigue. Also, our contacts noted that 'once is enough', meaning that not much difference was to be expected.

The following observations seem relevant, where we should keep in mind that no firm conclusions should be based on this low response rate:

- Although the response rates were low, overall, for all requirements and for all target groups, scores were consistent and slightly higher than for cycle 2.
- Scores were much higher for requirements that scored relatively low during the previous round: sharing information and self-healing.

We take these outcomes as a confirmation that CS-AWARE was well received, in general, and that it technically performed according to the requirements. The improvements made as a result of user feedback were appreciated as well. The users now appreciated and understood self-healing.

5.7.2 Evaluation of Usability in Cycle 3

For the final test, we prepared a set of use cases for participants to resolve. As explained above, because we were testing self-healing, we used a test environment, and not the live system of the municipalities. We checked, in addition, that self-healing also worked in the target systems, so our usability test would be valid for the actual environment as well. We will discuss the outcomes for usability for the main screens of the console.

Figure 5.10 shows an example of the opening screen for the usability test. We asked users to resolve 4 different threats. The threats for which a question mark is displayed in the 'state' column have self-healing options. The colours indicate the severity of a threat, in the order red-orange-yellow-blue, and green. Green threats are the outcome of a keyword search from relevant social media. The example screen and the user actions refer to 'threat perception', one of the affordances of CS-AWARE.

For the usability test, the following use cases were presented for participants to resolve, for each participant in a different order, where two of the cases involved self-healing:

1. Social media report: during the previous cycle, we implemented the option for the system to engage in search in relevant social media on the basis of user-defined keywords. We asked the participants to interpret and handle an outcome of a search, as well as to feed a new keyword.
2. Vulnerability pattern: a software vulnerability may put systems at risk
3. Suspicious behaviour: a system behaviour monitoring pattern is triggered that indicates suspicious behaviour
4. General Security warning: a possible Denial of Service attack

The opening screen (Fig. 5.10) displayed all of these threats (and other) at once, and we asked users to focus on the four specific threats, one by one. For Larissa, the participants were four system administrators, for Rome the main participant was the database department manager, who assigned tasks to two system administrators, who also participated in the test. Both these situations were realistic. Table 5.21 summarises the usability final test.

Fig. 5.10 The opening screen used in cycle 3

Table 5.21 Main features of usability tests in cycle 3

Summary	Usability study, Rome-Larissa, cycle 3 FINAL TEST
Use cases	Social media report Vulnerability case Suspicious behaviour General security warning
Components	All components of the system are involved, we had specific interest in self-healing.
Participants (Rome)	2 system administrators (SUET), 1 manager (head of the data Center of Roma Capitale), 1 system administrator
Participants (Larissa)	4 system administrators, specialised in applications or in database services
Method	Cognitive walkthrough
Outcomes	All participants passed this final test

All users immediately inspected in the table (in Fig. 5.10 on the right) the type of threat and the system component affected.

This clearly was acquired behaviour we did not find during our first testing, or not to be carried out as efficiently. Clicking in the table in the row of the threat (except when clicking in the 'where' column, see below) will take the participants to the threat description window (Fig. 5.11).

This window is the core screen from where to act. It describes the threat, at a general level, the observed data (technical details), the current course of action and the action that were already undertaken. Furthermore, it allows the participant to comment on a current state or action, to assign the threat to someone else (ticketing, as an outcome of cycle 1), and to mark a threat as resolved, when applicable.

For users, the comprehension of the threat and undertaking appropriate mitigation actions are done from this window. For example, a user can decide to accept the self-healing suggestion and apply it, as has been the case in the example in Fig. 5.12.

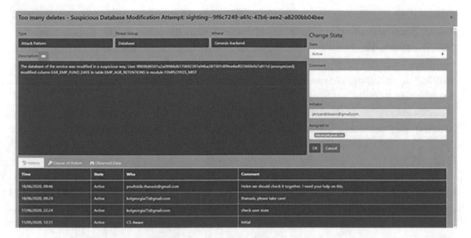

Fig. 5.11 Threat description Window

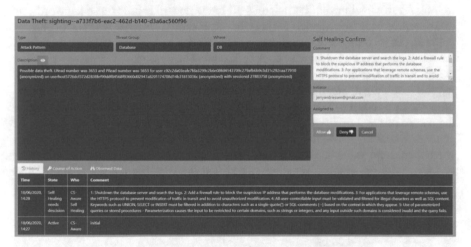

Fig. 5.12 Threat Description window with self-healing suggestion

This can be done quickly, or with ample reflection, or through assigning it to one or more experts within the user group. Participants had no problems with these possibilities, their main reflections involved thinking about what to comment about their chosen action. All had learned (from the previous cycle) to inspect and interpret the details of the observed data.

The participants in our usability test acted very effectively, especially by assigning any threat to another expert when this was seen as necessary.

When a self-healing suggestion is available, it is displayed in this screen (Fig. 5.12). While, during cycle 1, participants were hesitant to accept the suggestion, and information on the mitigation actions involved in self-healing was not as complete as in the current version, the final test showed more appreciation of self-healing and its implications. Users appreciated that this process was not automatic: there was the possibility to accept or reject the self-healing option, and the additional possibility to comment and decide if the threat had been resolved or if additional work was required. User agency in self-healing appeared to be an important asset. Participants were able to apply self-healing, often after consultation of the relevant experts.

Concerning projection of a threat, to study possible impact of a threat in the municipality network, CS-AWARE offers the network visualization, which (since cycle 2) immediately focuses on the part of the network implied by the threat (Fig. 5.13).

Many things are shown in this screen. First, there is the network component that may be compromised, as well as the system nodes that are linked and possibly in danger. Also, the other threats are displayed. The threat that is in focus, and its main information can be seen in the right-side window. There (infoflow) it can be seen in what work processes the implied node is involved. Finally, we would like to mention the 'keywords' option, where users can insert keywords they think are relevant to look for in social media. Participants all knew how to handle these keywords, as these were already investigated during cycle 2. It should be noted that not all participants inspected the system visualization. Threats can be resolved without looking at the visualisation, and this was sometimes what happened. This means that the description of a threat in the respective window was often sufficient for understanding and resolving a threat. But we can also conclude that handling of an emergency often took precedence over learning and deeper understanding. This was clearly due the nature of the usability test, but as well to the nature of their job description in general. Explicitly engaging into learning and increasing awareness takes place when there is no immediate threat and a user has time (allowed by their manager…) to 'play' with the system. We did not test that directly, nor did the usability and awareness questionnaires (2 and 3). This is an organizational matter as well: is

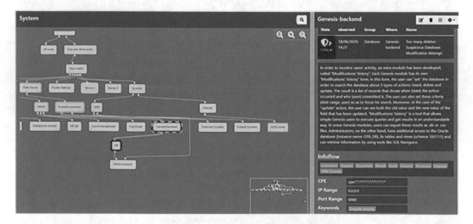

Fig. 5.13 System visualisation (left) and threat description (right)

development of awareness seen as incidental or is it an explicit and desirable feature of professional activity?

On the basis of the current usability test, we can however safely conclude that all users had mastered the system and were able to perform all essential activities efficiently and flawlessly. This clearly was not the case during cycle 1. The users have learned, and the system has improved.

These results were confirmed by the quantitative outcomes. Questionnaire 2 was filled in by the users in Rome and Larissa immediately after the usability test. Table 5.22 displays the outcomes. The following comments can be made:

- Overall, the scores are somewhat higher than for cycle 2, which we take as an indication that the improvements were appreciated.
- Moreover, we can assume that users have advanced in their appropriation of the system, an idea which is supported by the qualitative analysis of usability in the previous section.
- Two questions have mixed scores, one is about moving back and forth between screens, which can be taken in different ways. Users who often want to check their ideas often go back to a previous screen. However, this may also signify lack of experience and understanding. So, the question is ambiguous, but still received high scores (signifying infrequent movement between screens) in cycle 3. The other question is about the opinion of the users that CS-AWARE requires experience to be used, which received mixed scores. Here, as well, mixed interpretations are possible. This implies that our measure for flexibility (of the system: a component of Operability) is unreliable.

The usability sessions, as well as the reliable handling of all feedback from users, have contributed much to the appreciation and appropriation of the CS-AWARE tool by the users.

Table 5.22 Outcomes for Questionnaire 2 for cycle 3 (for cycle 2 between brackets), for each requirement, proportion positive (0–1). Experience of the user (1:low −5:high)

Usability, CYCLE 3	Rome	Larissa
N	3	4
Experience	2	5
S9: Knowability		
Clarity	0.82 (0.67)	0.97 (0.85)
Consistency	0.73 (0.76)	0.98 (0.85)
Memorability	0.69 (0.69)	0.82 (0.72)
Helpfulness	0.8 (0.5)	0.70 (0.70)
S9: Operability		
Completeness	0.84 (0.78)	0.88 (0.73)
Precision	0.87 (0.78)	0.83 (0.72)
Internationalisation (S13)	0.8 (0.67)	0.85 (0.85)
Flexibility	0.53 (0.47)	0.45 (0.55)
Mean	0.76 (0.66)	0.81 (0.75)

5.7.3 Evaluation of Awareness in Cycle 3

Questionnaire 3 was filled in by the users in Rome and Larissa immediately after the usability test. Table 5.23 displays the outcomes. The following comments can be made:

- Overall, scores are somewhat higher than for cycle 2, which we take as an indication that the improvements were appreciated, and of increased appropriation.
- Scores for awareness of the phases of decision-making were very high. This can be interpreted as high user agency for resolving threats, as a result of greater awareness of this process.
- The relatively new assets of the CS-AWARE approach, sharing information and self-healing, receive good scores. We interpret that as users indicating improved awareness of sharing information and self-healing.
- The high scores for comprehension indicate that the information that the CS-AWARE system provides to the users is very well received.
- Participants of questionnaire 3 attributed very high scores to their awareness. This awareness pertains especially to the decision-making process when there is a threat.

5.7.4 Evaluation of the Organizational Level in Cycle 3

Questionnaire 4 was returned by 8 managers from Rome, and 1 from Larissa. Table 5.24 shows the outcomes. These are almost identical to the scores from cycle 2.

Because this level deals with managers, we asked the deployment teams to provide feedback on the 'look-and-feel' of the output table of CS-AWARE. This table is an excel file, that can be downloaded from the 'closed threats' page. This table has

Table 5.23 Outcomes for Questionnaire 3 in cycle 3 (for cycle 2 between brackets), for each requirement, in KPI (0–1). Experience of the user (1: low −5: high)

Awareness, CYCLE 3	Rome	Larissa
N	3	4
Experience	2	5
S1: Provide cybersecurity awareness		
Perception	0.80 (0.73)	0.90 (0.72)
Comprehension	0.79 (0.72)	0.95 (0.84)
Projection	0.75 (0.67)	0.95 (0.85)
Decision	0.8 (0.76)	0.91 (0.76)
S2: Sharing		
Sharing	0.7 (0.54)	0.86 (0.80)
S3: Self-healing		
Self-healing	0.71 (0.72)	0.89 (0.63)
Mean	0.72 (0.65)	0.89 (0.74)

Table 5.24 Outcomes for Questionnaire 4 in cycle 3 (cycle 2 scores between brackets), for each requirement, in KPI (0–1). Exp means experience of the user (1–5)

Organisation, CYCLE 3	Rome	Larissa
N	8 (12)	1 (5)
Exp	1.8	1
S10: Compliance with internal regulations		
Perception	0.80 (0.80)	0.88 (0.76)
S10: Service delivery		
Comprehension	0.76 (0.74)	0.84 (0.74)
S1: Provide cybersecurity awareness (in organisation)		
Projection	0.83 (0.79)	0.80 (0.72)
Mean	0.79 (0.78)	0.84 (0.74)

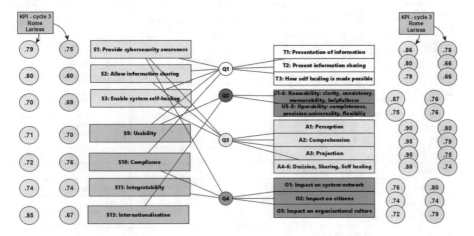

Fig. 5.14 Cycle 3: requirements and KPIs

the overview of the threats that were closed over a period of 1 month. It contains the following slots: closed at, first observed, group, Stix type, where, name, and description. We got replies from both municipalities that these were OK. This means that, for now, the main desired artefact (deployment scenario, slot 3, Table 5.2 and 5.3), from which many tables can be constructed, is satisfactory for the managers.

5.7.5 Conclusions After the Evaluation of Cycle 3

Achievements in cycle 3: Table 5.25 provides an overview of the cycle 3 outcomes. Figure 5.14 provides an overview of all requirements and scores for cycle 3. As can be seen, no score is below 70%, and most scores are above 80%. In terms of these scores, we can say that CS-AWARE has achieved all requirements in terms of KPIs.

The main achievement for the CS-AWARE system was the successful implementation of specific self-healing policies to the Rome and Larissa monitoring patterns.

Table 5.25 Summary of outcomes of cycle 3 for the two LPAs

Deployment	Rome	Larissa
Technical achievements	Instantiation of Roma and Larissa specific self-healing policies for monitoring patterns finished.	
Depl. Team	13 participants	6 participants
Objectives	Efficient detection, better collaboration	Effective detection
Foreseen impact	Improved communication with service providers	Learning, reputation
Desired	Alerting, reporting	Reporting
Evaluation	Rome	Larissa
Level 1: Technical		
Method	Questionnaire 1	
Participants	1 sysadmin, 4 managers, 4 service users	3 sysadmin, 1 service user, 1 manager
Mean score	0.78	0.89
Level 2: Usability		
Method	Cognitive walkthrough	Cognitive walkthrough
Participants	3 (1 data center manager and 2 sysadmin)	4 (system admin)
Outcomes	Final test passed	Final test passed
Processes	Distributed	Collaborative
Method	Questionnaire 2	
Participants	2 sysadmin, 1 manager	4 sysadmin
Mean score	0.76	0.81
Level 3 awareness		
Method	Questionnaire 3	
Participants	2 sysadmin, 1 manager	4 (sysadmin)
Mean score	0.72	0.89
Level 4: Organisation		
Method	Questionnaire 4	
Participants	12 (managers)	1 manager
Mean score	0.79	0.84
Level 5: Business		
Method	Questionnaire 5	
Participants	None	None
Scores	NA	NA

Also, we can say that users successfully completed the four exercises that were part of the final usability test. This means that our users, and very probably also users in other municipalities and professional contexts, with similar motivation and experience, can work with the CS-AWARE system to perceive, comprehend, project and mitigate cybersecurity threats. Moreover, we have shown that working with CS-AWARE increases awareness of cybersecurity, for individual users, and also within their organisation. This is the basis for realising the objectives that users formulated in the deployment scenario: efficient detection of cyber-threats, and better communication and collaboration between departments, between system administrators and managers, and between departments and citizen users of services. Better collaboration, learning, and increased reputation can be built on these achievements.

Lessons Learnt: The impact of the corona-virus during this cycle was twofold. In the first place, the number of questionnaires returned was about half compared to cycle 2. Of course, this could also be due to the impression by the pilots having to do the same thing twice, with no apparent changes. This explanation may hold, as the scores were the same or even higher than for the previous cycle of piloting. Especially self-healing and sharing information had higher scores. A final reason for lower return of questionnaire could be the very tight time frame. As we aimed for a final test, we wanted the questionnaires be filled in after the final test, which gave the respondents about a week for sending back the questionnaires, which might have been too short. Nevertheless, in spite of the low number of questionnaires returned, the scores were not affected.

Implications: We think the main implication is that our work in cycle 3 confirms very clear benefits of the CS-AWARE system for the pilot municipalities. They now have a cybersecurity system that detects and explains cybersecurity threats, which makes them more aware of cybersecurity, their system and the state of their organisation. In the next section, we will answer the main research questions formulated in Sect. 5.4 of this chapter.

5.8   Final Conclusions of Piloting

5.8.1   To What Extent Is the Technical Implementation of Cs-Aware Effective?

From the technical perspective, piloting results have shown that the general framework presented in Chap. 4 works as intended, and the technical functionality of each component is given in the generic case, as well as in the two pilot specific examples in Roma and Larissa. We have not identified any shortcomings or misconceptions of the technical framework that would necessitate a major revision of the implementation. This has been confirmed by the functional testing, as well as in the usability sessions during pilot testing. The positive outcomes have been confirmed by the evaluated results of questionnaires 1 and 2 in piloting rounds 2 and 3, with all KPI indicators showing a positive result. The analysis at the organizational level (represented by questionnaire 4) has not revealed any technologically relevant challenges that need to be addressed by the CS-AWARE framework.

Through the usability sessions with the users in Rome and Larissa we could confirm that the awareness created through CS-AWARE, and visualized to the user through the CS-AWARE user interface, provides insights into the security relevant system behaviour, and provides the user with information that is currently not available through any other technical means. A clear benefit and potential for efficient reaction to and mitigation of security issues has been identified. Those results have been confirmed through the results of questionnaire 3, with all relevant KPI indicators showing positive results.

One aspect that worked remarkably well is the interaction between workshop participants during the socio-technical system analysis (according to the methodology specified in Chap. 2), which has proven to be an exceptionally positive element in the creation of individual and organizational cybersecurity awareness. Throughout the workshop the users gained a deep technical understanding of their own systems and their security relevance, which simplifies the interactions with the technical part of the CS-AWARE system, since incidents based on behaviour monitoring detected by CS-AWARE are instantly comprehended and understood by users that helped to define them. Furthermore, since this analysis leads to system monitoring that the actual users and administrators of the system actually care about, an increased motivation to address potential issues detected by the CS-AWARE system could be observed. This is an observation that we could only capture qualitatively through our design-based evaluation method, and is impossible to quantify through KPIs.

It was shown that the two advanced concepts implemented by the CS-AWARE interface, self-healing and information sharing have been received well by the users and the potential for inclusion of those functionalities in the day-to-day work flow has been confirmed. It was observed for both functionalities that, since those concepts are relatively new and have not been available in any way before CS-AWARE in the two piloting municipalities, the time to get familiar with the relevant interfaces provided by CS-AWARE has been longer than for the other tested functionalities. This observation is confirmed by the analysis of results from questionnaire 1, with the KPIs for self-healing and information sharing initially scoring lower in round 2, but increasing in round 3. One aspect that cannot be quantified by KPIs but is a significant result relates to information sharing. It was brought forward by the municipalities that information sharing is not seen as a major technological challenge, and the functionality provided by CS-AWARE fulfils the technological need. However, sharing of security relevant information with parties outside the organization provides significant organizational and policy challenges that need to be solved first and are currently not addressed. It was acknowledged however that due to the reporting obligations of the NIS directive and the GDPR, municipalities are currently under pressure to get the relevant policies in place to address security relevant information sharing.

To conclude the analysis of the technical perspective of piloting results we would like to note that piloting has shown that the implemented CS-AWARE user interface concept is very well received by the users, and can be used to its full ex-tent after a short time by new users due to its focus on simplicity and efficient work flow. Most usability improvement suggestions have been implemented, based on feedback and observations during the usability tests and helped to further refine and significantly improve the workflow of the user interactions. Piloting has shown that the general framework works as intended and expected, and the technical functionality of each component is given in the general case, as well as in the two pilot specific examples in Roma and Larissa.

5.8.2 To What Extent Is the Cs-Aware System Usable by Expected Target Users?

From the perspective of usability, the users attributed very high scores to this aspect of the CS-AWARE system, and scores increased over time, which implies that they appreciated the improvements to the system that were made on the basis of their feedback and the outcomes of tests.

CS-AWARE is not a simple system to use. This is especially due to the complexity of information for cyberthreats, the difficulty of comprehending this information, understanding the role of compromised software or technology in the context of the municipal network, and making the correct decision for threat mitigation and protection of the network and the citizens. Such is the work of professionals, and they cannot make mistakes without consequences for the network, departments in the municipality, and, ultimately, the citizens.

We have deployed and tested the system in the context of two municipalities, with very different ways of working. In the municipality of Larissa, system administrators performed all the tasks, or consulted with their direct colleagues. In Rome, expertise was distributed, and various tasks were handled by different persons.

For each of these tasks, the system provides options. Users indicated that:

- The elements presented by the interface were clear, and their structure and functions were clear
- The elements were presented in a consistent manner
- The way the system works is easy to remember
- Working with the system does not require much help, with some time for learning it
- The presented information is complete, for making decisions
- The descriptions are clear and precise
- The system allows flexible use, users can go back and forth as they please, and retract most earlier steps or decisions
- The system presents the information in the language of the user, translations work well
- The options that the system presents are sufficient for acting

Moreover, the learning curve for using this complex system requires a couple of hours of practice, perhaps in the form of a guided session with an experienced coach, for example a system administrator who has already worked with the CS-AWARE system.

For new users understanding how to work with the system, we highly recommend usability sessions with new and beginning users. This could be part of new user sites co-designing and appropriating the system through socio-technical workshops.

To conclude, the CS-AWARE interface supports immediate detection of threats (depending on the cycle for scanning the municipality network), and supports comprehension of the nature of the threat, the part of the system network that is compromised and the other nodes that may be involved, and offers understandable self-healing suggestions when these are available.

5.8.3   To What Extent Is the Awareness of Users Affected by Discussing and Using the System during Deployment?

First, we note that the awareness of cybersecurity was already positively affected and enhanced as soon as the project started, and most significantly for users participating in the SSM workshops (Chap. 2). For the purpose of evaluation, we started interpreting awareness after the story workshops at the beginning of year 2 (Chap. 3), and again at the start of deployment at the beginning of year three.

We defined awareness as a concept with 6 components: knowledge of cybersecurity threats (1), the system network (2), the organisation (3), external cybersecurity-related organisations and communities (4) as well cybersecurity agency: knowing how to act in case of a threat (5), and acting when there is no threat (6).

Agency can be defined here as the possibility for actively contributing to cyber-security. It alludes to the capacity of humans to distance themselves from their immediate surroundings and it implies recognition of the possibility to intervene in, and transform the meaning of situated activities (Mäkitalo, 2016). With our qualitative approach (stories, deployment scenario), we could interpret organisational and communal awareness, as well as agency when there is no threat. With the usability exercises (also qualitative), we could investigate user agency when dealing with threats, and, to some extent, how users exploited the affordances of the CS-AWARE tool for comprehension, also in the context of their system network, and for sharing with external organisations. Finally, with the awareness questionnaire, we could check how users rated their use of the affordances of the CS-AWARE tool during the phases of decision-making in case of a threat. We assumed that higher ratings for use signify greater awareness.

Concerning knowledge about threats (1), we can say that instead of laborious sessions inspecting logs and internet sources, looking for information that would lead to diagnosis of a threat, as happened before CS-AWARE, this is now handled by the CS-AWARE system, and in such a way, that information about threats is readily and immediately available for the user. This means that potentially, threat awareness is increased. Users agree with this statement, reflected by a score for requirement S1 (provide cybersecurity awareness) that is very good. We therefore conclude that knowledge of threats is increased with every new threat, especially when users are sharing their experience with others, during or after threat resolution.

The knowledge by users of their municipal system network (2) was already greatly fostered during the SSM workshops (Chap. 2), taking place before deployment. We have evidence that the awareness ratings for system visualisation and projection of threats have gone up during deployment, and we can therefore conclude that also system network awareness is addressed in a more than suffi-cient manner.

The organisational level of awareness is addressed below, through evaluation question 4. The level of external organisations, communities, or other departments involved in cybersecurity (awareness component 4), who all would have an interest in sharing information about cyberincidents, as it stands now, is merely a technical asset. The concept of sharing information involves knowing what to share, and with whom, for what reason. Our participants were aware of the need for sharing infor-mation in general, but there was no installed policy for this, neither was there explicit mention of some form of collaboration with other communities or depart-ments in this respect. Both municipalities were aware that this issue needs follow-up.

We understand most about the awareness component of user agency for handling cyberthreats (5). When we say that CS-AWARE is an expert system, we crucially refer to the process of dealing with threats. This process was different in the two pilot municipalities. In Larissa, the system administration department handled all cyberthreats. This is a small department, with experts who can work together to resolve a threat. The way they currently work does not have to fundamentally change with CS-AWARE. In Rome, expertise on all aspects of the extended network of nodes and services, is highly distributed. Handling cybersecurity threats requires a central expert who delegates different tasks to different experts, who are respon-sible for their particular system or service. More often than not, handling a cyber-threat will involve more than two system administrators who work in different departments. For the communication that this process requires, Rome already had a ticketing system in place. CS-AWARE is made to work in both contexts, so a ticket-ing system was implemented as well. These differences have implications for awareness and on how expertise is distributed between users.

We distinguished four main phases in the process of dealing with threats, and users indicated (in the questionnaires) good awareness of all of these phases. From the usability tests, we learned a number of specific things for awareness:

- Concerning perception, the opening screen of CS-AWARE provides immediate awareness for those involved in monitoring.
- Concerning comprehension, we noted that users attend to the main characteris-tics of a threat (type, date, system component involved), but not always to all details (detailed description, system information and history). This may have good reasons, linked to a user's expertise, and the need for immediate resolution may require efficient handover. However, we observed that threat comprehen-sion has more attention from those who are responsible for all aspects of threat resolution, than from those who deal with some part of that process.
- Concerning projection, the same applies as for comprehension. While some users study the network visualisation extensively, others do not look at it, and focus on their own 'section' of the network. It should be noted that the actual repair, by inspecting log files of the affected system component, still takes place 'outside' of CS-AWARE, except in the case of self-healing. We highly recom-

mend training for new users to focus on projection of threats through system visualisation with CS-AWARE.

- Concerning decision-making, we noted that handover of threat mitigation to other users was the rule rather than an exception. Also, we noted that most users have the habit of checking if their decision was implemented correctly (e.g. threat now listed in resolved threats, or handover now included in current threats). It was clear that this already was part of their normal routines, and now made explicit (and recorded) through CS-AWARE.

As a conclusion for agency in handling threats, as a component of cybersecurity awareness, we can say that CS-AWARE greatly facilitates user agency, making detection and mitigation more efficient and effective, with the additional asset of better comprehension and projection of threats. The extent to which comprehension and projection abilities of users increases, depends on the extent to which users pay attention to this information. The very positive outcome of the questionnaire seems to be relative to the roles and actions of users during threat mitigation.

Finally, how about user agency when there is no immediate threat (6)? We know that system administrators in Larissa have been 'playing' with CS-AWARE when there were no threats to resolve. Of course, this is an important activity for gaining experience. System Administrators in Larissa stated their ambition for learning in the deployment scenario. They had an interest in an improved reputation for their department, as a consequence of improved services. This could lead to considerations about weaknesses in the network components and development of new services for citizens. On the other hand, in Rome, ambitions were formulated for the management level, in terms of more and more effective interactions between departments in the context of cybersecurity. Although the managers in Rome were very positive about possible organisational impact (see the next section), it remains to be seen how these expectations will be realised.

Concluding, through deployment of the CS-AWARE system, we can say that cybersecurity awareness in both pilots has been greatly increased, at the level of threat detection and mitigation, and, to a somewhat lesser extent, to understanding and learning about threats, also in the context of the system network. Further work is expected for the exploitation of increased organisational awareness in Rome, and the elaboration of sharing threat information with relevant external agencies and authorities.

5.8.4   *To What Extent Does Using the Cs-Aware System Have Impact on Cybersecurity Awareness at the Organisational Level?*

As has been previously set out, system requirement S1 of the CS-AWARE system is to raise and maintain awareness of cybersecurity both at an operational and organisational level. During cycles 1, 2 & 3 the users completed questionnaires that covered a range of topics, each of which included potential aspects of awareness. The responses we received to questionnaire 3 in cycles 2 & 3 enabled us to gauge the growth of the respondents' cybersecurity awareness as the project progressed, the base-line having been established by interpreting the outcomes of the story telling workshops. In terms of the organisational effects of the CS-AWARE system, questionnaire 4 established the extent to which it impacted upon organisational culture and security awareness.

It is clear from the responses to questionnaires 3 & 4 that the system has, in the opinion of the respondents, significantly improved both their cybersecurity awareness and that of the organisation(s) in which they work. However, and despite positive responses concerning organisational impact, we are concerned about the issue of maintaining cybersecurity awareness at the organisational level of an enterprise over a period of time.

We might surmise that as those staffs that were initially involved in the SSM workshops and in the consequent adoption of a CS-AWARE system move onto to other roles – possibly in other organisations, the state of "organisational awareness" of cybersecurity in an organisation may begin to decay. This is of course both a problem of knowledge management and of organisational memory – the issue of how an organisation maintains its knowledge base in despite the inevitability of staff turnover.

In our view this can be avoided in three ways: Firstly, by instituting an annual SSM workshop involving personnel from both systems admin and senior management to review and if necessary, re-tune and update their organisation's CS-AWARE system – ideally as part of an IT audit in compliance with ISO 20701. Secondly, ensuring that a formal management reporting line is established that provides at least quarterly reports concerning the effective operation of the CS-AWARE system and outputs from it to senior management at board level of the organisation. Thirdly, by using the CS-AWARE system console to maintain a record of mission-critical security incidents and of how and the extent to which the consequences of these incidents were mitigated.

References

Abildgaard, S. J. J., & Christensen, B. T. (2017). Cross-cultural and user-centered design thinking in a global organization: A collaborative case analysis. *She Ji: The Journal of Design, Economics, and Innovation, 3*(4), 277–289. https://doi.org/10.1016/j.sheji.2018.02.003

Alonso-Ríos, D., Vázquez-García, A., Mosqueira-Rey, E., & Moret-Bonillo, V. (2009). Usability: A critical analysis and a taxonomy. *International Journal of Human-Computer Interaction, 26*(1), 53–74. https://doi.org/10.1080/10447310903025552

Ambrosino, M. A., Andriessen, J., Annunziata, V., De Santo, M., Luciano, C., Pardijs, M., Pirozzi, D., & Santangelo, G. (2018). Protection and preservation of Campania cultural heritage engaging local communities via the use of open data. In *DGO, 21st Annual International Conference* (pp. 1–8). https://doi.org/10.1145/3209281.3209347

Andriessen, J., & Pardijs, M. (2021). Awareness of cybersecurity: Implications for learning for future citizens. In Z. Kubincová, L. Lancia, E. Popescu, M. Nakayama, V. Scarano, &

A. B. Gil (Eds.), *Methodologies and Intelligent Systems for Technology Enhanced Learning, 10th International Conference. Workshops* (Vol. 1236, pp. 241–248). Springer International Publishing. https://doi.org/10.1007/978-3-030-52287-2_24

Anttila, J., & Knowledgist, V. (2006). General Managerial Tools for business-integrated information security management. In *Applied information technology research: articles by cooperative science network. Lapin yliopisto* (p. 8).

Barab, S., & Squire, K. (2004). Design-based research: Putting a stake in the ground. *Journal of the Learning Sciences, 13*(1), 1–14. https://doi.org/10.1207/s15327809jls1301_1

Berliner, D. C. (2002). Comment: Educational research:The hardest science of all. *Educational Researcher, 31*(8), 18–20. https://doi.org/10.3102/0013189X031008018

Engeström, Y. (2011). From design experiments to formative interventions. *Theory & Psychology, 21*(5), 598–628. https://doi.org/10.1177/0959354311419252

Eyvindson, K., Kangas, A., Hujala, T., & Leskinen, P. (2015). Likert versus Q-approaches in survey methodologies: Discrepancies in results with same respondents. *Quality & Quantity, 49*(2), 509–522. https://doi.org/10.1007/s11135-014-0006-y

Furnell, S., & Vasileiou, I. (2019). Chapter 1: A holistic view of cybersecurity education requirements. In I. Vasileiou & S. Furnell (Eds.), *Cybersecurity education for awareness and compliance* (pp. 1–18). IGI Global). https://doi.org/10.4018/978-1-5225-7847-5

Hibshi, H., Breaux, T. D., Riaz, M., & Williams, L. (2016). A grounded analysis of experts' decision-making during security assessments. *Journal of Cybersecurity, 2*(2), 147–163. https://doi.org/10.1093/cybsec/tyw010

Janhunen, K. (2012). A comparison of Likert-type rating and visually-aided rating in a simple moral judgment experiment. *Quality & Quantity, 46*(5), 1471–1477. https://doi.org/10.1007/s11135-011-9461-x

Mahatody, T., Sagar, M., & Kolski, C. (2010). State of the art on the cognitive walkthrough method, its variants and evolutions. *Intl. Journal of Human–Computer Interaction, 26*, 741–785. https://doi.org/10.1080/10447311003781409

Mäkitalo, Å. (2016). On the notion of agency in studies of interaction and learning. *Learning, Culture and Social Interaction, 10*, 64–67. https://doi.org/10.1016/j.lcsi.2016.07.003

McKennel, A. C. (1974). Surveying attitude structures: A discussion of principles and procedures. *Quality and Quantity, 7*(2), 203–294.

NIST. (2018). *Framework for improving critical infrastructure cybersecurity, version 1.1* (NIST CSWP 04162018; p. NIST CSWP 04162018). National Institute of Standards and Technology. https://doi.org/10.6028/NIST.CSWP.04162018.

Paavola, S., Lakkala, M., Muukkonen, H., Kosonen, K., & Karlgren, K. (2011). The roles and uses of design principles for developing the trialogical approach on learning. *Research in Learning Technology, 19*(3). https://doi.org/10.3402/rlt.v19i3.17112

Piccolo, L. S. G., & Pereira, R. (2019). Culture-based artefacts to inform ICT design: Foundations and practice. *AI & SOCIETY, 34*(3), 437–453. https://doi.org/10.1007/s00146-017-0743-2

Schønheyder, J. F., & Nordby, K. (2018). The use and evolution of design methods in professional design practice. *Design Studies, 58*, 36–62. https://doi.org/10.1016/j.destud.2018.04.001

Strauss, A. (1988). The articulation of project work: An organizational process. *The Sociological Quarterly, 29*(2), 163–178.

Strode, D. E. (2016). A dependency taxonomy for agile software development projects. *Information Systems Frontiers, 18*(1), 23–46. https://doi.org/10.1007/s10796-015-9574-1

The Design-Based Research Collective. (2003). Design-based research: An emerging paradigm for educational inquiry. *Educational Researcher, 32*(1), 5–8. https://doi.org/10.3102/0013189X032001005

Thomas, M. K., Shyjka, A., Kumm, S., & Gjomemo, R. (2019). Educational design research for the development of a collectible card game for cybersecurity learning. *Journal of Formative Design in Learning.* https://doi.org/10.1007/s41686-019-00027-0

Wieringa, R. (2014). Empirical research methods for technology validation: Scaling up to practice. *Journal of Systems and Software, 95*, 19–31. https://doi.org/10.1016/j.jss.2013.11.1097

Chapter 6
Cybersecurity Awareness in Rome and Larissa: Before, During and After CS-AWARE

Arianna Bertollini, Massimo Ferrarelli, Omar Parente, Claudio Ferilli, Thanasis Poultsidis, Jerry Andriessen, Thomas Schaberreiter, and Alexandros Papanikolaou

6.1 Introduction

Thanks to the expansion of digital technologies, municipalities are becoming intelligent, allowing the interconnection of systems, people and devices to improve infrastructure, efficiency and advantage for residents. Many local authorities have started investing in smart skills and are evaluating how they can leverage them to improve services and reduce costs. However, the benefits of technology can also pose dangers for townships. The use of Internet-connected systems and the offer of online services increase their vulnerability to a cyber-attack.

Roma Capitale and the municipality of Larissa are part of this context, which in their digitization process feel the need to face the problem of adequately securing their digital services, systems and networks. Hence their adhesion to the European project CS-AWARE to collaborate for a cyber-security platform development.

A. Bertollini · M. Ferrarelli (✉) · O. Parente · C. Ferilli
Municipality of Rome, Rome, Italy
e-mail: arianna.bertollini@comune.roma.it; massimo.ferrarelli@comune.roma.it; omar.parente@comune.roma.it; claudioguido.ferilli@comune.roma.it

T. Poultsidis
Municipality of Larissa, Larissa, Greece

J. Andriessen
Wise & Munro, The Hague, The Netherlands

T. Schaberreiter
CS-AWARE Corporation, Tallinn, Estonia
e-mail: thomas.schaberreiter@cs-aware.com

A. Papanikolaou
INNOSEC, Thessaloniki, Greece
e-mail: a.papanikolaou@innosec.gr

© The Author(s), under exclusive license to Springer Nature Switzerland AG 2022   145
J. Andriessen et al. (eds.), *Cybersecurity Awareness*, Advances in Information
Security 88, https://doi.org/10.1007/978-3-031-04227-0_6

In this chapter the two municipalities report the experience and evaluation of their participation in CS-AWARE and discuss how the project has changed their awareness of cyber security, the impacts on their internal organization by their collaboration in team building activities.

6.2 How the Chapter Was Written

The chapter is conceived as a personal report produced by authors from the two municipalities, as a response to a set of structuring questions by the editors. What they wrote is included as personal sections, embedded in more explanatory sections where the editors provide background to the user experiences. Everything that is reported has been voiced by participants from the two LPAs. The information provided about the local context, the starting situation, ambitions, and impact are derived from minutes and reports of the CS-AWARE workshop sessions and team meetings in which they participated. Especially, the information that the users provided during a story telling workshop and during a deployment scenario workshop will be exploited. We briefly describe the context of these two workshops here.

The purpose of the story telling workshop was for the CS-AWARE project to get insight into experiences of the users (the system administrators, managers and citizen-users) about cybersecurity issues before the project. We wanted to obtain a sketch of typical cases of cybersecurity dangers, and the needs, roles and issues of people dealing with these dangers in their professional contexts. This was a necessary asset for our project, because no technology solution can be successful without considering user needs, in addition to involving the users in the design of the technology. We decided that the best way to capture experiences was in the form of a story. The story telling workshop involved small groups of participants from the municipality (system administrators and users from various departments) in collaborative efforts to produce meaningful stories about their cybersecurity experiences. The workshops were organised both in Larissa and Rome at the beginning of the second year of the (3 year-) project. The procedure was generally as follows: Before these workshops, we asked participants through email to think about a personal experience with cybersecurity. During the workshop itself, users were asked to group into small teams that set out to deepen the content of these experiences according to a limited set of topics: the role of the organisation, the role of the system, the actions that were undertaken for resolution of the issue, the emotions of the user, and a conclusion providing perspective, e.g. lessons learnt, or things that could change. Then, we asked a representative of the group to present the story to all present. In addition to the project capturing these user needs, achieved by an analysis of the stories produced, the way we set up the story telling workshop also had a clear impact on the awareness of the users themselves, as they will recount below. We will report some examples of stories and some analysis outcomes that illustrate the state of cybersecurity in the two municipalities.

A year later, when the CS-AWARE technology was ready for deployment in the municipal contexts, we organised a deployment workshop. Users, together with researchers of CS-AWARE, worked for a full day on the creation of a deployment scenario, by collaborating in small groups. The main thrust of the workshop was to articulate and share local goals and expectations for deployment of the technology, on a shorter (during deployment) and longer (after deployment) term. This information was collected immediately after the third of the series of SSM workshops (Chap. 2), during which users were involved in discussing their network architecture, define critical processes, and symptoms of critical use cases, activities which were crucial for the design of the technology. The deployment workshop captured users' perceptions and expectations before actual deployment. The participation of the users in this workshop also served to raise their awareness about cybersecurity issues, and about what they expected of the technology.

To reiterate, the information in this chapter is based on outcomes of the story telling workshop (beginning of year 2), the deployment workshop (before deployment at the beginning of year 3), and on contributions by users (from the IT-departments of Rome and Larissa) written at the end of the project for the current chapter.

6.3   The Experience in Larissa

The CS-AWARE project for Larissa began due to the very active EU-Projects department of the municipality. Our colleagues informed the IT department that a project called CS-AWARE was searching for pilot cities. After a short investigation, the IT team lost no time and in good collaboration with other members of the consortium, Larissa was in. That was the first step.

The project proposal was accepted. The project kick-off meeting was announced, and one member of the IT department was sent to Oulu to meet the other partners and to find out what exactly the project was about. The kick-off meeting was quite revealing, and outlined the challenging requirements for participation of the piloting municipalities in the project. With tight time limits (only 1 month), the Larissa team then had to prepare the conditions and settings for the first SSM workshop. The Municipality welcomed the CS-AWARE analysis team and five full days of information exchange were devoted to the analysis of the municipality's systems. The Soft System Methodology was used in order to guide the amount of information. The IT department of the municipality, despite their deep and complete knowledge of its systems, learnt that it did not have a very high degree of sensitivity about security.

The next few months work was done on the clarifications of the network components and their role in the system. The CS-AWARE framework was ready, and we had to adopt it. This implied that we had to identify the critical points in our system. We had to write down the processes that include sensitive and critical information for the municipality. The CS-AWARE team worked on these tasks during the second workshop. In the next few weeks, the first testing application was installed in our systems and data was sent to the CS-AWARE repository for processing. The time had come to search into all these data and identify patterns. Suspicious behaviour, that our departments want to be aware of, since it indicates unusual or malicious acts. During the third workshop long discussions took place about patterns. Also, a big part of the third workshop was about the deployment plan.

Meeting by meeting, workshop by workshop, the project was evolving, and the IT department's security awareness was following, too.

6.4 Added Complexity for Rome

The same series of activities were organised with Rome, but in a much more complex context. To deal with the workshop's organization a project manager has been appointed and a very heterogeneous working group has been set up. Colleagues with specific skills in terms of security were brought together with colleagues with skills related to European projects, IT skills, skills related to the management of economic resources. Altogether it was necessary to require and to assemble a wide variety of skills and competences, all within the Municipality of Rome. Furthermore, given the structural and organizational complexity of the municipal context, the participation of other staff inside and outside the administration was sometimes required, on the basis of issues and needs that gradually arose. It is important to point out that during the project we experienced various kinds of complexity: from administrative to technical, from organizational to cultural.

The workgroup—more or less stable due to resource turnover – was in charge of coordinating various activities, handling administrative matters, organizing workshops, participating to live and virtual meetings, presenting the project during several thematic events, updating citizenship on activity progress using Roma Capitale's web portal. The planning and the realization of a single travel or of a local event had to face a new approach, based on the confrontation with a variety of internal stakeholders and external partners.

Participating to CS-AWARE project was a great challenge for all of us, not without encountering difficulties mainly due to the fact that we were not immediately aware what final results would have been. Each of the involved stakeholders, including system providers, system administrators, managers, had to face a more flexible and unstructured environment, in which the uncertainty about the results played a central role. This was not an ordinary, routine job: we had to deal with real innovation!

During first and second workshops, several application and infrastructure areas contact persons were involved to provide the necessary overview of our complex administration's functioning. In the subsequent workshops, due to other project partners requests, areas of interests were selected and studied in more depth, creating models using Soft System Methodology approach.

6.5 Cybersecurity Awareness in Rome and Larissa Before the Project

The chance to participate in the European CS-AWARE project was perceived by Roma Capitale as an opportunity. Security in cyberspace is today one of the main needs for those working to guarantee community interests not only by ensuring a proper reliability of their digital services, but also by doing that in the most secure manner, in order to protect their data. Security's culture spreading is a must: both security and privacy must be adequately taken into account in all management decisions and actively lived: at the moment, the digital services security-by- design principle is not yet an integral part of our organizational culture especially if we consider it from our internal users point of view which still remains largely a "paper-based" one (but things are changing, maybe faster than expected!).

The municipality of Rome does have a centrally organized and managed network of computers, software, databases, and services in a dedicated IT Department, but, at the beginning of the CS-AWARE pilot project, it did not have an effective organization to implement a coordinated cybersecurity policy yet. The municipality is equipped with heterogeneous systems, not always interconnected, which give selective types of information to selected group of users only. Several providers work on logic, hardware, software, and perimeter security quite independently. Cybersecurity was a separate concern for each different provider. This is one of the reasons why different roles and objectives quite never match each other. We did not expect this situation to change by the sole participation in CS-AWARE, but we aimed for increased awareness on potential benefits for Roma Capitale's organisation.

A problematic aspect present before the project was a lack of awareness of the risks related to cyber security by most employees of the organisation, starting from the daily actions up to the management of external attacks. One of the most common duties worked on by employees is e-mail messages management which, even if it may seem trivial, hides many pitfalls due to lack of knowledge on dangerous messages and threats that can lead to security impairment of the systems. Another aspect relates to application services access management through weak passwords chosen by users who tend to select simple and plain words sometimes base on their personal interests or environment.

A story may illustrate this point:

> One of the tasks of employee E1 is to read several hundreds of emails every day. Despite strict regulations, E1 opened an infected email. This mail had no subject, which could have caused suspicion. After E1 opened the mail, files in each directory on his computer he accessed became inaccessible. The IT-Department was notified, they discovered a type of ransomware that had encrypted all the files in the directories E1 had accessed. The computer was restored in the IT-Department, where all files that were not yet accessed were retrieved, and the system was reinstalled, and the intact files were restored. Infected files could not be retrieved. E1 was initially terrified, feared the computer was lost, but now looks at it as a lesson to apply the rules more strictly for such cases.

The story is interesting because we should be aware there are persons for whom it is a duty to open all email. It also shows that these and other users may be very careful, but still occasional mistakes can slip through. The third interesting point is the role of the IT-Department: they solve the issues, but they also have strict rules. This may lead to users feeling insecure about the IT-issues they still might have.

So, a cybersecurity project aimed to enforce the awareness of the matter is fundamental to ensure in this way both security and privacy do become an integral part of the administration's culture path towards a complete digitalization. In addition to that, an effective data security policy is an absolute need in order to allow personalized services and data interoperability between our internal department and/or with external administrations, even more in a smart city perspective.

In Larissa, there is one system administration department for the municipality (including smaller municipalities around Larissa), where all cybersecurity issues are handled. System administrators in Larissa have different expertise, are aware of each other's competences, and regularly share information. Managers from other departments seem or need to be less involved. In Rome, there are many departments responsible for many services, making for a complex structure. Departments handling data or applications, or citizen internet services, do not necessarily share information on cybersecurity. For awareness of cybersecurity, this means the information is highly distributed, and often unshared.

Developing more knowledge on cybersecurity is a local affair, and this may work fine in Larissa, but less so in Rome. Application of safety rules is monitored in Larissa by the system administration, although it is less clear if other department managers share this attitude. In Rome, if and how safety regulations are monitored, is highly dependent on the department managers. The role of system administrators is often incorrectly perceived by non-technical users to help them anyway, sometimes even helping them to avoid the general regulations, in order to allow them to make their own work.

Here's another story: some users are responsible for managing European contracts. The problem that is frustrating them, and probably many others, is that some websites from the EU are forbidden by system administration. The procedure for employees is for the head of the department to write an official request for access to certain websites by certain people. This helps, but only for the people who have been listed in the request. All of this is interpreted as erratic by our user: is there any real policy behind this?

The issue seems to be that the policy behind allowing or restricting websites within the municipality is unclear, and probably erratic, in the sense that it may be the result of many policies operating at the same time, including individual decisions. Consequently, some things are allowed, and some things are not allowed. For the user, this is highly unsatisfactory and confirms a lack of trust, in addition to searching for individual solutions rather than for those that relate policy issues in general. There is clearly a transparency problem both for the IT-Department AND for individual users.

The rules and norms that the IT-Department maintains underlying the regulations for monitoring safety behaviour, and providing services to the users, are different in the two municipalities. In Larissa, we see that the system administration expects citizens (internal users) to stick to the rules and regulations. Conversely, the citizens expect an immaculate and timely resolution of their cyber-issues. This means there is always a tension: trust may grow but also decrease. In Rome, there seems to be a general norm that the system is too complex, and we should accept imperfection to the extent that not all policy regulations are created in the interest of maximal cyber-security. It looks like the various IT-departments are still able to provide quick services to users with problems. These services are provided on an individual basis, for a user who detects an issue, other users are not necessarily informed in case of a security issue. There is a lack of shared awareness and culture about threats, especially in terms of understanding the threat and its possible impact on the system network. In Rome, this lack of awareness is related to the distributed knowledge of the complex system network: there is no single individual who completely oversees the whole network.

Concerning sharing of information within the organization, in case of Rome, there is the issue of information sharing and knowledge management: knowledge is highly distributed and less shared, system users seem to have no concern for each other's cybersecurity issues, and the culture favours general regulations but individual solutions (not shared). For awareness, this creates a problem: it does not evolve nor spread. Also, in Larissa, while employees within the system department are aware of the cyberincidents that have occurred, this knowledge remains within the department, and seems of less concern for others, although the other employees that we met seemed to have some interest.

So, what can we assume about the status of cybersecurity awareness in the two municipalities before the start of the project? Awareness can refer to knowledge, but also to (the possibility for) action. In CS-AWARE, we think cybersecurity awareness originates in sharing and collaborating for understanding, not only a single cyber-incident, but also the context, the causes, and the possible impact. Although the situation in Rome reflects a more complex organisation for handling cybersecurity, underlining the importance for managing and sharing information, the main characteristics of how cybersecurity threats were handled in Rome and Larissa were (perhaps surprisingly) similar.

The perception of threats is not immediate, usually the system administrator is informed of an incident (by a service user, or a system alert), and some harm may already have been inflicted. Then, the system administrator must identify the threat, discover its possible impact, such as inferring when the issue probably appeared, in what part of the municipality network, or with which employees. Our impression is that the focus in both municipalities is on threat identification, and less on other processes that are part of awareness: threat comprehension, understanding of impact in the system network and its business processes, and decision making. We will explain this below. So, perception is not immediate, identification can be time-consuming, mitigation focuses on individual service users, and is local (part of the system), and not (holistic) system oriented.

For further understanding of the possible impact of a threat, if it is not already well known, a system administrator must access further resources for identification.

Searching for and reading relevant and reliable information on the internet takes time, depending on the experience of the user, and often does not happen at all. The user may have to research log-files of the system, and maybe of services, to figure out what exactly is the problem. System administrators may do this, if they are responsible for the network or a part of it. If their role is to manage the database of a service, they may not be interested in such log files. In that case, communication with another specialist is needed, which requires information about the threat to be collected and understood.

For awareness this means that for a system administrator resolving a threat, full comprehension is not always necessary, and sharing information is only needed sometimes. Therefore, comprehension is limited, at the individual and at the organizational levels.

If the system administrator wants to understand the risks imposed by a threat to a system node, as well as to the other nodes, and which services will be in danger, a thorough understanding of the network as well as of the process in which the nodes are involved is a requirement (see Chap. 1). As awareness often is distributed, these aspects of resolution are a problem for users, especially in larger organisations. Users may be very knowledgeable for their own service but may not always oversee the possible implications for the rest of the network.

For awareness, this means that projection is local, and users may lack the knowledge to research their network. In Larissa, communicating with colleagues often solves the problem. In Rome, such communication may be more complicated, and requires the involvement of management and issue ticketing systems.

The final step in threat resolution is decision-making, and this involves several different actions. In the baseline situation, it usually means resolving the threat, for example by applying a patch or update, or blocking a user or part of the network. In a complex organisation, it may mean referring resolution to the expert responsible for that part or service in the network. In all these situations, experts, colleagues, including managers, may be informed, or consulted. This requires sufficient comprehension. Also, the user may need to ascertain that a threat is resolved. Most awareness in the base-level situation depends on communication between different stakeholders. In other words, awareness requires the collaborative attitude of equality of contributions, desire for sharing, and careful consideration of contributions by all.

6.6   Expectations of Using CS-AWARE in Rome and Larissa

Roma Capitale and Larissa represent public administration of different size and complexity, but with the same challenge: to ensure the security of their growing number of data, which belongs to the entire population of their citizens, now and even more in the future, allowing a secure storage and exchange, preserving them from malicious or direct attacks. The data integrity and security are a prerequisite to protect their value and to share it among their citizens, also empowering the competitiveness of the cities in an increasingly data-driven economy.

The expectations of three groups of stakeholders (system administrators, managers, and local users) were collected before deployment of the software. Among other things, we addressed their objectives, what they expected the system to do, and how they expected their own behaviour to change concerning cybersecurity. Participants had all been involved in the SSM design workshops, so they were already quite aware of the CS-AWARE affordances.

System administrators from Rome, currently people overseeing their own part of the municipal network and software, expected easy threat identification, preferably a trouble ticket on their mobile device, in case a threat was assigned to them. With that, they would like detailed information about a threat: type, time of incident, and all other information for better understanding of the threat. This would result in faster and more effective threat resolution. They would then expect to use the tool on a daily basis, and also be engaged in more reflection on past problems and solutions, and generally in more regular communication within the technical team and with internal users. Monitoring of incidents, and assigning (ticketing) threat resolution in Rome is and remains a management job, involving several people.

System administrators from Larissa, working as a team covering most of the municipality network, were interested in timely detection of threats, and absence of false alarms. They stressed the importance of trustworthiness of the information provided by CS-AWARE, and this information being up-to-date. The insisted on the system allowing them to decide, rather than relying on automatic system choices. They focused on learning by reflection and discussion. They expected to assign monitoring roles between them.

At the management level the differences between Rome and Larissa are very clear. The managers in Rome, who were directly involved in the workshops and design team meetings, focus on better collaboration within the municipality, between managers of the various departments, between managers and service providers, and with senior managers and users within the municipality. All of this may be realised on the basis of weekly incident reports and monthly trend reports, better quality of services and therefore more trust with the general public. There is clear awareness that this requires extensive discussions amongst policy makers.

The managers in Larissa were less involved in the workshops, or not at all. They expect the reputation of the information department to improve, and therefore general trust in system administration.

Finally, the users of the network and its services have an ambiguous role. On the one hand, they expect cybersecurity to operate without their awareness, and without their involvement. They simply want their system to be safe and reliable. Users in Rome were not interested in notifications about threats, especially not before they were resolved. In general, and this is the common situation probably everywhere, users are supposed to follow regulations. Users in Larissa said they appreciated some information about the threats that were resolved.

6.7   Outcomes

The main outcomes from the perspective of the municipalities includes aspects relating to increased reflection on cybersecurity issues in the organization, an increased understanding of the socio-technical structure of their organization, and the benefits from the increased internal team building and collaboration efforts within the organization and with academics. Those aspects will be discussed in this section.

6.7.1   Increased Reflection

It was clear that participation in project meetings, workshops and internal reflections in the context of CS-AWARE already had a great impact.

From the very beginning of the CS-AWARE project in Larissa, the initial task was to identify the persons to be involved from technical, users and managers perspective. All of them were about to be key players to the development, implementation and support of the project. After the work group was established, roles were assigned and the project could start for Larissa. The soft systems methodology was utilized for identifying the critical and sensitive points that the system should care about. Although the IT departments had deep knowledge about their infrastructure and software, never before were they triggered to spot the weak and sensitive points. Therefore, the next workshops proved very useful and training-like for the IT. The whole process changed the point of view of IT employees in Larissa, and created an improved security perspective about their systems. Companies and universities were directly in touch with the IT personnel and influenced towards raising the security awareness in Larissa. The collaboration between Larissa and the technical experts, security consultants and academics was constant and went both ways; both parts benefited from this part of the project, resulting to a better focused approach and final implementation specifically in Larissa.

In Rome, as the project produced its first usable results (i.e. graphical interface first demo in the second year of the project), we clarified, in the local deployment team, our ideas on what, till then, had only been abstract descriptions, concepts and questions becoming gradually simpler to understand and provide answers to different requests. This way of collaboration forced all of us, each for her/his part, to make an effort to identify elements that sometimes seemed obvious but actually are not. Furthermore, especially during workshops—when each contact person illustrated the part of her/his competence—different participants had the opportunity to listen and understand descriptions of activities, processes, area's difficulties different from their own getting an opportunity to know better the internal structure of the administration. Having such an overview will surely facilitate each of us in carrying out daily activities. All of this was even better understood by those who had access to the CS-AWARE web platform prototype, especially on system section tab where anyone can immediately get an idea of our local administration complex infrastructure's network, and, even more important, in asking critical questions, such as: How does a denial service attack may impact our organization? Or: How can we build effective knowledge management for collecting and comparing social media reports about new threats?

6.7.2 Increased Understanding of the Internal Organization

Roma Capitale Digital Transformation Department's participation in the project certainly generated a greater awareness of cybersecurity in all employees, in particular it stimulated the curiosity of those colleagues who, by function, never had the opportunity to address these issues and provided an opportunity for deepening to all those who, on the contrary, were already aware of it, by work mission. Our experience in CS-AWARE project has shown us the importance of disseminating and growth of awareness of cybersecurity in our organization, not only in the ICT department, but in all the central departments and territorial structures ("Municipi", the 15 municipalities in which Roma Capitale is subdivided). A possible organizational approach in order to spread and cultivate the new cybersecurity culture might be to use the network of the local contact points for the digital transformation ("referenti della trasformazione digitale"): such a network, formed by 50–60 people with at least basic ICT skills and competencies, is already currently under training about the implementation of a new "digital workplace" concept, which has made mandatory after the massive use of remote working due to the CoViD19 emergency. Thus, we can suppose that also cybersecurity could play a crucial role in their training programs and in their hypothetical day-by-day operational mix in the articulate organization of Roma Capitale.

6.7.3 Teambuilding and Internal Collaboration

In Rome, the cross-cutting nature of the CS-AWARE project imposed a multi stakeholder approach which led the creation of a teambuilding practice never experienced before on these matters. The need to meet, discuss, create working groups with colleagues and technology's suppliers, relate to an international environment and collaborating with the municipality of Larissa, have certainly represented an enrichment for all of us both from a professional and human point of view.

A team made up of people from different backgrounds, subject matter, and mentality in their way of working is difficult to work with a smooth flow, especially if several members are participating for the first time in such a group. In the case of Larissa, a group of private and public employees was set up, without having taken part in such a large-scale program. Specifically, employees of the municipality of Larissa with the role of administrator in different computer systems of the municipality, suppliers who were invited to work for the first time in a public body in a European program, as well as people who were invited to work with the municipality of Larissa to improve security according to instructions from senior government agencies, they were key members of the Larissa team.

In Larissa, the cooperation of private sector employees with employees of the municipality in the framework of a European program on security, was an unprecedented undertaking which at least initially had some degree of difficulty. Until the start of the CS-AWARE program, the municipality of Larissa had experience with external partners, but only at the level of procurement or service. Never before have employees been asked to work with one of the municipality's associates to run a pilot project at European level. The difference in the objectives of work, the distance between them, the chronological deadline in cooperation with the other members of the program, and in general the difference in the way of working between the public and the private sector were a peculiarity which all parties were called to overcome in order to ensure the smooth cooperation between them.

In parallel with the program, the municipality of Larissa, following instructions from the Cyber Security Directorate of the Ministry of Digital Governance under the guidance of the European Union, was invited to introduce new GDPR rules for the management of the municipality's computing data. In this the managers of the computer systems of the municipality, who are also employees of the municipality, were invited to cooperate with a new DPO (data protection officer) from the private sector which in fact had no experience with any form of public body. The new DPO, of course, joined the Larissa team for CS-AWARE. Thus, in parallel with the project, the employees had to work with a new external partner, without knowledge in the computer structure of the municipality and the operation of the structure, in order to create and integrate these new security rules. This was an obstacle, since for several steps in order to complete the deliveries for CS-AWARE, the employees had to first consult the DPO, taking into account the distance and the different objects of engagement. Nevertheless, the cooperation with the members of the Larissa team went quite smoothly resulting in the good operation of the team.

6.7.4 Collaboration with Academics

Computer and network security is an increasingly important and complex branch of IT, especially with the new capabilities of computers and new forms of data sharing, processing and storage. Thus, understanding the concepts of security and techniques used to maintain the safe operation of computer systems in the municipality and other bodies is now of great importance. The academic offer in this case is very helpful, since the concepts and techniques of security are clarified and understood, which when applied in practice often are different from their theoretical approach.

The academic contribution has been of great importance and very useful throughout the project. The different approach, more theoretical and according to academic standards, helped to better understand the concepts related to security but also in general to the understanding of the entire computer security sector. The employees of the municipality of Larissa found quite useful the guidance given to them by the members of the project who have the experience in such large-scale cooperation in European projects. Applying security practices in day to day work, depending on the work objective of each employee, requires the knowledge of the proper tools but also the broader theoretical background that the academic approach can offer. After the project, now, computer security and problem solving in this area are more familiar in both theoretical and practical background, and are more effective in solving them following a multi-faceted approach - theoretical and practical.

6.8   Perspectives for the Future

Without adequate security protocols and monitoring systems, civic authority's systems can be easily exploited by hackers by taking control of computer servers causing possible theft of personal data. To keep up with the cybercrime constantly evolving threats and tactics, municipalities like Roma Capitale must be proactive, not reactive, on cyber security. All this has led to an internal re-organization in order to create specific divisions designed to monitor and contrast any attacks with the support of technology.

One possible re-organizational scenario could be using dedicated people (adequately trained) for CS-AWARE platform monitoring able to manage different threats by themselves or by using system advice, maybe forwarding tasks to selected teams.

The CS-AWARE platform acquisition together with organizational re-engineering could bring benefits such as (but not limited to) easy identification and classification of the threat with decision support system for damage prevention; improved quality of the service which led to more satisfied citizens and personal data protection, hence increased trust in the LPA.

The CS-AWARE project brought several positive results for the Municipality of Larisa. Initially, it resulted in the creation of a tool that warns users of the application about security risks on the municipality's computer network. A tool useful for monitoring the safe operation of the municipality's computer systems and crucial in detecting electronic threats, as well as for informing any similar threats that had been dealt with in the past in one of the pilot municipalities. Equally important, however, are the experiences gained from the employees during the project and from the cooperation with the team members, scattered in different regions of the European Union. These experiences involve both technical knowledge of the security of computer systems and networks, and how a program of this scope works in collaboration with its members.

The way a team of a European program operates is very different from the way in which the employees of the municipality of Larissa work on a daily basis. The number of members involved in the program, as well as the different subjects, the different working conditions, the different cognitive backgrounds that led to different approaches to certain issues and the different ways of dealing with the problems presented were unprecedented for the employees who participated for the first time in a project with so many participants from different places and with different work objectives, especially when these members came from places outside Greece with completely different mentalities.

This gave the employees the experience of working with universities and academics, members and shareholders of large companies and employees of a metropolitan municipality, that of Rome, which is very different from the municipality of Larissa. The behaviour of the other members in the discussions and in the adoption of the way they will work during CS-AWARE, their way of thinking, their acceptability in some cases where there were small difficulties that had to be addressed, or the establishment of common solutions to overcome controversies have left a fairly positive impression on officials. The way of thinking, which is very different from the one that was approached so far by various issues as well as the flexibility in adapting to certain situations, in combination with the above, are characteristics that the members of the Larissa team are willing to adopt in future collaborations.

Even for employees who had no background in computer science in general, the CS-AWARE project and their contact with its platform helped them understand security as a concept and why it is so important in the smooth running of the Municipality of Larissa. Understanding the concept of "awareness" is quite important in avoiding any infections from various threats, which users who have simply been asked to evaluate the ease of use of the application are now able to understand thanks to this project.

6.9 Aftermath

There are several ways in which the CS-AWARE system provides the user with possibilities for increasing awareness. Most of these possibilities are part of the threat detection and resolution process. More awareness by users could lead to changes in awareness by others in the organisation and to organisation change, if so desired, but obviously this is not an automatic consequence of more awareness.

Some of the main affordances of the CS-AWARE system are:

- It provides automatic threat detection and identification, and additional information about the threat from reliable internet sources
- It provides contextualised information by visualising the threat in the system network components and business processes
- If available, it provides suggestions for self-healing, that can be applied automatically (when authorisation is granted)
- It allows automatic sharing of cybersecurity information with cybersecurity authorities

This report of the municipalities (largely written by the municipalities) shows evidence of increased awareness linked to these affordances: threat detection, comprehension of dangers of threats, understanding of possible impact on the local network, understanding of the network itself, more collaboration and communication within and between departments, more awareness of the human factors in cybersecurity, and the awareness that much more needs to be done.

What any technology will not (and should not) be able to handle is the impact of awareness at the organisational levels. For cybersecurity as a collaborative activity of all stakeholders, communication between various departments is essential, as well as sharing of relevant information, and, crucially, a policy for knowledge management. While cybersecurity awareness may reveal the need for all of this, implementation of further steps can only be a slow process. CS-AWARE has given this process a big boost, which is all that we can hope for. Cybersecurity awareness is not a state, but an ever-ongoing process involving collaboration between all stakeholders.

Chapter 7
Marketing a cybersecurity Awareness Solution in LPA Contexts

John Forrester, Manuel Leiva Lopez, and Massimo Della Valentina

7.1 Introduction

Marketing to the public sector is already a formidable task compared to marketing in the private sector. There are numerous rules, regulations, priorities and challenges to be taken into consideration. A one-size-fits-all approach to government marketing does not exist given the differences in services supplied and agendas on the local level. This chapter will deal with marketing cyber security solutions in Italy in the public sector and will focus on municipalities.

Issues like cloud computing, cybersecurity, and broadband connectivity are all issues of interest to all levels of government. Cloud computing, in particular, is a priority to government agencies, from local to national level. Some regions have even created their own clouds and in-house software firms to help manage their work. Cybersecurity follows closely as a priority with an-increasing awareness of ransomware and the vulnerability of cloud storage. Local governments understand that broadband and Internet connectivity could potentially transform the lives of individual citizens and the community in general. Obstacles remain for anyone interested in selling cybersecurity to the public sector with the lack of resources in local government and the limited availability of qualified staff making it difficult.

The budget of local governments is largely determined by population. Average population size of municipalities in Europe (Table 7.1) show that except for the Netherlands, Eire and UK, large areas of other European countries are dominated by small and medium sized municipalities who typically are suffering from a scissor's crisis with increasing demand for new and innovative social services and declining

J. Forrester (✉) · M. Leiva Lopez · M. Della Valentina
Cesviter Consulting, Rome, Italy
e-mail: j.forrester@cesviterconsulting.it; m.leiva@cesviterconsulting.it;
m.dellavalentina@cesviterconsulting.it

© The Author(s), under exclusive license to Springer Nature Switzerland AG 2022   161
J. Andriessen et al. (eds.), *Cybersecurity Awareness*, Advances in Information
Security 88, https://doi.org/10.1007/978-3-031-04227-0_7

Table 7.1 Average population size of municipalities in Europe (2013)

	Number of Municipalities	Population	Average Number of inhabitants per municipality
United Kingdom	419	65,648,054	156,678
The Netherlands	390	17,064,682	43,755
Ireland	126	4,786,562	37,988
Poland	2479	38,131,648	15,382
Italy	7958	60,589,445	7614
France	36,658	65,129,822	1776
Finland	313	5,534,655	17,682
Austria	2100	8,745,151	4164
Germany	11,313	82,220,424	7267
Greece	325	11,149,330	34,305
Denmark	98	5,745,874	58,631

budget along with few qualified employees capable of handling new innovative services like cybersecurity software.

National governments in Europe have in recent years devolved a number of powers and responsibilities to local governments. In addition, there has been in many areas a push for more unions between local authorities to use more effectively the available resources. Indeed, in France, municipal cooperation is a crucial part of the national government strategy, considering the large number of "micro-municipalities". The Council of European Municipalities and Regions (CEMR) in its 2016 report on local and regional governments in Europe (Martinez Marias et al., 2016) that 60% of the decisions taken by municipalities and regions are affected by European legislation and almost 70% of public sector investments in Europe come from local and regional governments. Clearly these figures demonstrate the important role of local and regional governments in Europe's economy. They are a vital factor in the continuing social and economic transformation of European society. Local governments have very often key players in dealing with and overcoming the series of challenges that Europe has been encountering in recent years.

7.2   The Case of Italian Municipalities

Italy's economy comprises a highly developed industrial north, dominated by private companies, and a less-developed, highly subsidized, agricultural south, with a legacy of unemployment and underdevelopment. The Italian economy is driven largely by the manufacture of high-quality consumer goods produced by small and medium-sized enterprises, many of whom are family-owned (comprising over 99% of Italian businesses and producing around 68% of the national GDP). Italy also has a sizable underground economy, which some estimate accounts for as much as 17% of GDP. These activities are most common within the agriculture, construction, and service sectors.

It should be remembered that despite the existence of the EU common market, the cultivation and maintenance of personal relationships are still a vital part of doing business in Italy. Finding the right local agent, distributor, or business partner remains essential to enter the Italian market. It is rarely effective to rely on agents or distributors in neighbouring markets. At least within the LPA sector, this issue is not that uncommon in other countries in Europe. In small and medium-sized LPAs across Europe, experience has shown that most municipal leaders prefer to deal with local or regional contacts (OECD, 2009, p. 7).

The Italian cybersecurity market is largely driven by investments of larger companies, where most of the top management has become more aware of the increasing risk of intrusion into their business information systems. Italian firms are also becoming more concerned about threats to data confidentiality, integrity, availability, and authentication. The financial and utility sectors are generally the top end-users of ICT security in Italy, followed by defense, national and local government, manufacturing, transportation, and sectors. Medium-sized companies, which are a significant portion of Italian enterprises, and to a lesser extent smaller companies, are increasingly in recent years investing in security. However, there remains resistance on the managerial level to expenditures for cybersecurity security activities due to the perception in both business and government sectors that security is a cost rather than an investment.

In recent years, the government is starting to recognize cybersecurity as a national priority. In February 2016 the Italian government presented the first-ever National Cyber Security Framework (NCSF) which was developed from the Framework for Improving Critical Infrastructure Cybersecurity compiled by the U.S. National Institute of Standards and Technology (NIST).[1] The voluntary framework was meant to serve as a common frame of reference to identify existing and future standards and regulations. According to the Cyber Intelligence and Information Security Centre of the Sapienza University in Rome, city governments, local units of the National Healthcare System, and public hospitals are still the most vulnerable to cybersecurity threats.

Larger local governments (regions like Lombardia and the bigger cities like Torino) increasingly perceive cybersecurity as a core business requirement and security spending has been growing and will certainly continue to grow across the board, particularly in those areas which need to be improved to reduce vulnerability. The most important market drivers for cybersecurity include (OECD, 2019; US Department of Commerce, 2016):

- Increased security awareness and interest in complying with current legislation regarding cybersecurity;
- Challenges arising from the adoption of new technologies and business models requiring the implementation of security measures such as secure mobility and virtualization;

---

[1] https://clusit.it/pubblicazioni/.

- New government measures and investments to protect the digital identity of citizens and critical infrastructure from increasing cyber assaults;
- The implementation of Italian legislation calling for security measures to protect the privacy and personal data of citizens and compliance with national and international norms like EU General Data Protection Regulation (GDPR); and
- The implementation of industry-specific legislation, calling for compliance with national and international norms.

A common characteristic of all these market drivers is an increasing awareness in some select sectors of the importance of cybersecurity but not necessarily accompanied by a concerted effort to dedicate resources and staff to facilitate change.

A cybersecurity campaign, in our experience, could be developed in Italy focusing on a select number of regional and provincial institutions or with Unions of Municipalities, or possibly one of the Metropolitan areas. From experience in the field, bureaucracy in these Metropolitan areas pose significant problems for anyone interested in selling new and innovative software. In addition, the lines of authority and division of duties is not always as clear as they could be. The regions might be easier to investigate as they are often bettered resourced with clearer mandates regarding what they are responsible for. Each Metropolitan area in turn would involve their network of institutions. Italy, for instance, is dominated by small municipalities who by themselves cannot afford to pay much for software. Part of the issue is a question of priority. Faced with increasing demands for their already limited resources municipalities often do not perceive security as a pressing issue when they have many people out of work, threatened with the loss of their homes, and dealing with COVID-19. As we outlined below, finding those unions of municipalities or those that have cooperating agreements to operate jointly critical services like the police will probably be in the long run more profitable than going from one municipality to another. It is possible to use the public procurement authority like Italy's MEPA but you still need to find local governments or agencies that are interested in combined offers. Responding to requests of individual municipalities online in MEPA is still time consuming.

The strategy behind any eventual marketing activity could also follow a process of "indirect dissemination" through larger organizations and a selected number of government associations in each country. In 2020 in the CS-AWARE project, we worked with a number of local agencies on cybersecurity and the solutions offered by the project. Under the auspices of the CS-AWARE project we held a workshop in Caligari (Sardinia) in November 2019 and later in January 2020 in Barletta (Puglia) on cybersecurity. Those that attended were quite interested in what they could done. Unfortunately, the start of COVID-19 exerted a negative impact on the follow-up activities.

While the dissemination models will differ from country to country (and also within countries: north and south in Italy), depending on both the systems of government and the interests of the stakeholders, we had hoped to follow-up with other activities to show how an approach like that of the CS-AWARE was particularly suited to local government. Dealing with associations of municipalities, or provinces, or even regions could be a good use of time and energy. However, the degree in each country to which these national associations perceive cybersecurity as a critical priority varies a great deal. Further on we discuss how a great deal of preliminary work on content is needed to show prospective clients in government agencies why proper cybersecurity solutions to be implemented. Field work has shown that often public officials are forced to make decisions about cyber security plans based on budgetary constraints and little has been done to help those in smaller municipalities to develop creative non-traditional solutions that do not require large sums of money. More work needs to be done to develop the observations in Chap. 2 "The Socio-Technical Approach to Cyber-security Awareness", since it is a strategy that could be very persuasive to policy figures in LPA's. Municipal figures are often most interested in understanding how the "soft-systems" analysis aspects would better help them fulfil their basic duties of supplying and maintaining social services for their communities.

In Italy the National Association of Cities (ANCI) has been quite active on a national level working on issues of security in recent years. Unfortunately, the weakest link in the chain, those municipalities of less than 5000 inhabitants have been largely left out. Much has been said about the need for digitalization and more preparation regarding cyber security and how to develop adequate defences, but not enough political commitment has been demonstrated to help train employees and develop new procedures. Employees remain unprepared for the impact of digitalization and the increasing number of cyber security attacks. Many municipalities are unstaffed and still trying to deal with the effects of COVID-19. These observations come from field work in municipalities throughout Italy working on dissemination and marketing in the CS-AWARE project While there are exceptions, the smaller municipalities are generally left to fend for themselves. Some regions, like in Lazio (Latium), have been trying to help with extra funding and additional resources for municipalities impacted severely by the recent pandemic.

The total population of Italy is 58,983 (Population of Italy, ISTAT, 2022). The total number of municipalities in Italy is 7958 (Feb. 2018) The average resident population in Italian municipalities is 7614 inhabitants. Some 5541 municipalities have less than 5000 inhabitants; in other words, around 70% of Italian municipalities are very small in size. Typically, their budgets are limited and have few qualified staff available. The smaller municipalities often have to share their staff with other municipalities.[2] In recent years, many municipalities have begun pooling their resources trying to meet the increasing demands for public services.

Currently there are 537 active Municipal Unions (called: Unioni di Comuni) with 3095 municipal members, that is, approximately 39% of the total number of municipalities. Only 17 of these Unions have more than 100,000 inhabitants. 8 big municipal unions associate more than 20 municipal members. One target for the

---

[2] One of the key figures in the administration of a municipality is the segretario comunale (municipal secretary). Very often the smaller municipalities share this person with other municipalities to save on costs. In other places, we discovered that police and social services are, for instance, often shared.

cyber-security campaigns could be the 12 municipal Unions that have more than 16 municipalities as members. It is also possible to consider as a target the 96 municipalities that have the status of provincial capital.

Using the site for demographic statistics (http://www.comuniverso.it[3]) 14 metropolitan areas can be identified in Italy (Table 7.2). According to Wikipedia, Metropolitan areas usually refer to a region consisting of densely populated area and its surrounding territories, sharing industry, infrastructure, and housing. As social, economic, and political institutions have evolved over time, these metropolitan areas have become key economic and political regions.

The metropolitan city of Roma Capitale is the biggest area and given the number of municipalities (1274) and residents, a possible target for anyone interested in marketing cyber-security solutions to local government. One issue that we discovered working with Roma Capitale on the CS-AWARE project is that there are overlapping areas of responsibility in the metropolitan area of Rome so it can be difficult to find the most appropriate person to connect with. Other regions the organizational structure has developed in a more coherent fashion with clear lines of authority and responsibility.

Since most metropolitan areas include multiple jurisdictions and municipalities (note the 121 municipalities associated with Rome), the metropolitan areas in EU countries will be of clear interest in terms of dissemination and marketing

Table 7.2 The 14 Italian metropolitan areas

The 14 Metropolitan areas					
Region	Metropolitan City	Number of Communities	Surface (km²)	Population (2017)	Demographic Density (p/km²)
Lazio	Roma	121	5363	4,353,738	812
Lombardia	Milano	134	1576	3,218,201	2042
Campania	Napoli	92	1179	3,107,006	2635
Piemonte	Torino	316	6827	2,277,857	334
Sicilia	Palermo	82	5009	1,268,217	253
Puglia	Bari	41	3863	1,260,142	326
Sicilia	Catania	58	3574	1,113,303	312
Toscana	Firenze	42	3514	1,014,423	289
Emilia-Romagna	Bologna	55	3702	1,009,210	273
Veneto	Venezia	44	2473	854,275	345
Liguria	Genova	67	1834	850,071	464
Sicilia	Messina	108	3266	636,653	195
Calabria	Reggio Calabria	97	3210	553,861	173
Sardegna	Cagliari	17	1249	431,430	345

---

[3] This site was recently taken off-line. Hopefully it will be restored in the coming months. For years it was one of the most authoritative sources of demographic information about Italy.

possibilities. Typically, these metropolitan areas are better endowed with budgetary resources and qualified staff. Metropolitan areas in Italy are by law supposed to help the municipalities in their areas on "digitalization" projects. Issues of budget restrict what can be done but the technical departments in these Metropolitan areas do what they can to help out. The primary issue in selling a cybersecurity software solution is to provide adequate and informed content that shows not just that you have a superior solution and how it can help its final users achieve their goals.

Aside from metropolitan areas, the comuniverso.it site notes that there are approximately 90 municipalities that are not provincial capitals and have more than 50.000 inhabitants. Those municipalities with more than 50.000 inhabitants are probably the most likely to have the social and economic resources to be interested in cybersecurity solutions.

Most Italian municipalities, as noted before, are small and lack the resources to be able to manage a cybersecurity application by themselves. The situation does not change substantially in other EU countries. Most municipalities in the EU are small and medium sized in population with insufficient resources to deal with cyber security issues. In addition, we have seen over time that there is a serious issue with municipalities not licensing properly their computers and software. The consequences are that many are unnecessarily exposed with out-of-date copies of Windows and software like Office. Over the last 10 years people have attempted to adopt open-source software but due to a lack of resources nothing much on a national level was achieved.

There are three issues to consider before beginning to plan a marketing campaign. First, it is important to remember that potential customers in the public sector need considerable education and guidance to understand cybersecurity in general, the threats they face, and what solutions might help them. This is particularly true of "policy" figures in local government who are not generally well versed in cybersecurity and are often unaware of how cybersecurity should be a critical part of the general preparedness plan of any government agency. Secondly, many agencies do not prioritize cybersecurity. There is plenty of evidence to show that cybersecurity is vitally important for protecting data and privacy. Unfortunately, many do not give much importance to security issues until it is too late. The need to prioritize cybersecurity needs to be carefully explained without resorting, necessarily, to lists of potential dangers and terror tactics. Warnings of potential doom rarely sell much, no matter how well intended the message is. The final issue to keep in mind is in the public sector, in our experience, policy figures are not technically oriented, so it is far better to focus on needs and what your solution might do for their community rather than on strictly software and hardware involved (US Department of Commerce, 2016). We also discovered on the CS-AWARE project that solutions like CS-AWARE that emphasize soft systems analysis typically appeal far more to public officials who are concerned about understanding their needs and issues before committing to a series of hardware and software solutions that may or may not turn out to be durable solutions.

7.3   Marketing Strategies

The most effective strategy to use in the public sector is that of Inbound Marketing which is, generally speaking, a business methodology that seeks to draws in customers by creating valuable content and experiences tailored to them. While outbound marketing interrupts your audience with content they don't always want, inbound marketing seeks to create connections to what customers are looking for and tries to solve problems they already have. This is critical in the public sector where officials are responsible above all for furnishing a series of services to their communities so they are principally interested in how a cybersecurity solution can help them maintain a constant flow of services. It is important to remember that while each of the tactics that will be explored below can certainly be implemented independently, but they are far more compelling when used together. Together, they comprise the practice of what many call "inbound marketing". Originally developed by Hubspot, inbound marketing is the latest form of marketing, and it has proven to be generally successful. While traditional forms of marketing consist of bringing your product or services to potential customers and fighting for customer attention, inbound marketing is more about letting your customers find you.[4]

While inbound marketing is about drawing customers to your company, outbound marketing is about your business pushing its message out to the world, usually through advertising. Pushing your message in front of potential customers through outbound marketing can be costly. Even buying advertising on Google or similar is rarely cheap, and results can be elusive since consumers tend to ignore ads. We have found in Italy that few municipalities have employees trained in the use of social media and the use of search engines like Google, partially due to lack of resources and partially to a degree of scepticism about the returns they could hope for. In contrast, inbound marketing can be low cost or even free, and because you are forming relationships and links, you are also building a lifetime value of your customers.

This relationship building in the context of the public sector is best achieved through a variety of tactics and strategies. Examples would include whitepapers, video tutorials, webinars, podcasts, and social media posts. If the content you make available benefits the customer, they will be more motivated to buy. Particularly in the public sector, inbound marketing requires time to develop good content, nurture customer leads, and create "evangelists" for your company. Having evangelists for your products can be very important to building support for a company. Overtime they start to become the face and voice of the company's mission. They start to narrate compelling stories for your products, engage customers via various channels,

---

[4] HubSpot is an US based developer and marketer of software products for inbound marketing, sales, and customer service. It was founded in 2006 by Brian Halligan and Dhamesh Shah (website: https://hubspot.com). Particularly for those new in the marketing field Hubspot can be quite important since they offer a platform of marketing, sales, customer service, along with CRM software and, in addition, explanations of methodology, resources, and support.

and build a community of customers. While evangelists can be quite important when their product is unique and people generally perceive it as useful, companies should focus most of all on developing new and compelling value for their customers through reworked value propositions and enhanced customer experience.

Understanding customers is important to recognizing the security issues that affect the public sector the most. Creating carefully crafted buyer personas (representations of customers based on actual customer data and market research—is probably the best strategy for acquiring a deeper customer knowledge and marketing to each of them appropriately. A customer-based persona is a "semi-fictional archetype" that reproduces the most important traits of a large part of your audience using the data you have collected from user research and web analytics. Personas help identify and prioritize changes to your offering based on what your customers need the most.[5]

For instance, IT staff may follow social media, but they are not involved in the decision-making process, in terms of finding suppliers. If we are speaking of an antivirus below 50€/year they generally have the autonomy to choose an antivirus, but the costs of solutions like CS-AWARE platform are far too high for them to act in an autonomous fashion. It becomes an issue for the policy branch where the needs are quite different. The focus of any campaign needs to be the policy branch (mayors and municipal councils).

Ultimately, the marketing and selling aspects of any campaign must be aimed at the political side. The IT staff will be charged with making it work, to an extent, dealing with high-ranking managers or directors may incur some benefit in that chances are higher that the mayor and/or the city council will follow the advice of a manager or director. The budget is controlled by the mayor and the city council so any content from a cybersecurity campaign should be directed at their concerns. Technical explanations run the risk of being counterproductive. As noted above, soft system analysis and its focus on the environment and the needs of the community, as the CS-AWARE solution does is more understandable to those in policy branches. Our experience working on the CS-AWARE project is that they tend to think in terms of services for their communities so information regarding case studies and programs launched elsewhere tends to be better perceived.

As far as social media is concerned, it is rarely clear whether an investment in a social media presence in the market, other than providing a reference point for the company, will be effective enough to target the appropriate "decision-makers". At least in Italy the average age of a manager or director in local public administration is rather high and combined with an inconsistent use of social media by employees to promote and find elements for their relative jobs rather complicates the issue.

LPAs are just a little microcosm of the PA (Public Administration). The only huge difference is that in LPAs most of the decisions are politically driven. It means

---

[5] https://socialmediatoday.com/news/what-is-a-buyer-persona-and-why-is-it-important/507404/. One important point in this article is that it is not uncommon to have multiple buyer personas for a government agency or business. For instance, if a product needs to be approved by multiple people before making a purchase, each individual involved in that decision should have a separate persona.

that, if you want to sell, you need to have direct contacts. Networking is the key. Most importantly, finding the right network can be critical. In Italy we found that the obvious choices like National Association of Cities (ANCI—Associazione Nazionale Comuni Italiani) were not necessarily the best to work with. An example of promising networks would be the various associations of Municipal Secretaries. These are important figures in local government bodies and have an important impact on the running of their local government agencies.

The environment around most LPAs also has a strong territorial affiliation, in terms of politics. It is rare to find someone not politically aligned in an LPAs with the respective governing Party. Municipal officers are particularly concerned about providing for their surrounding communities. Any issues like cybersecurity must be framed in this context and linked where possible to local concerns.

In this context, if we want to underline the real difference between the overall public sector and the LPAs in the Italian market, it is its permeability. The general public sector (that runs from a Ministry to the Postal service) can be entered (commercially and not) through a transparent process, according to the "transparency rule".

Marketing wise, with LPAs you are obliged, as a seller, to be in a relationship with the political counterpart. Of course, everything is conducted within a fairly well-defined legal framework. Not a single LPA will be interested in buying a product if you do not have a relationship with the corresponding political elements and a good product.

In 2003, the Italian Public Administration e-Marketplace (MePA) was introduced: it is a procurement platform managed by Consip SpA on behalf of the Italian Ministry of Economy and Finance (MEF). The MePA is a digital market in which any PA can purchase goods and services offered by suppliers, for purchases below the Italian threshold (at the moment 150 thousand euros). It is open to qualified suppliers according to non-restrictive selection criteria. The whole process is digital, using a digital signature to ensure legal compliance and overall transparency of the process. It works just like a real market, as the same products can be found and sold by different suppliers at different prices, terms and conditions. Suppliers can decide on the geographical area in which the delivery of their products / services will take place. The site of MePA is https://www.mepa.it. The importance of MePA is, certainly, not be underestimated. The MePA platform is free, electronic billing is offered, offers suppliers easy access to a vast market, greater transparency for bidding process, and possibility of updating one's offerings at any time.

The rules that suppliers must follow to register and sell to the MePA are established by Consip according to the different product categories. MePA connects thousands of public bodies and suppliers distributed throughout the Italian territory, both centrally and locally. Registered purchasing administrations can use the two purchasing administrations. Consip also defines the qualification requirements and terms of service conditions. In this case, in addition to having to provide the service completely (information and contractual conditions included) in Italian, in order to be able to register as a supplier to the MEPA, the company must have a VAT number registered in the Italian Chamber of Commerce, for that it must open an office in Italy or it must be a company already present in Italy.

From 150.000€ or more you need to be a different type of supplier (partizione gara d'appalto) or participate in a public call (until the end of July 2021, after which the amount will be reduced to 40.000€).

Generally, there are two scenarios that follow:

- First scenario: direct selling below the threshold of 150.000. Usually quite fast and functional, you sell your product and that's that. You follow the marketing tactics and advice listed above, particularly, in relationship with LPAs.
- Second scenario: participating in a public call but in 99% of the time, they are hard (or almost impossible) to win. Proceedings are extremely bureaucratic, expensive, and long. Unfortunately, the bureaucracy is still geared largely to respecting the form rather than considering the needs.

When dealing with customers, you should always remember that your customers are not the cybersecurity experts—you are. That means that rather than offer them the solution, you think, they have always been looking for, you should take the opportunity to educate them about newer, better solutions that they themselves might not have thought about before. Take the time to understand the security needs of prospective customers in the public sector and offer them clear advice and insights based on their needs, it will help you establish your credibility, encourage customers to engage with you even more, and retain a base of loyal customers.

One tactic to employ would be to show how immediate solutions to security issues do not necessarily imply buying pieces of hardware and software. Time could be well spent analysing the current situation. Customers could be encouraged to study what they are doing currently and what suggestions employees might have to resolve problems. Very often in our experience working out with customers how their work could be better organized to respect better basic "cybersecurity" hygiene habits[6] might be an important first step. Only after really understanding the current situation can solutions be developed for the future and create security solutions that are flexible and capable of responding to changing circumstances.[7]

Applying an inbound methodology is great way to reinforce your message through the building of more lasting relationships with possible customers. It shifts the focus to valuing and empowering potential customers to achieve their goals at any stage in their journey with you. This is exactly why it is so critical to provide content that offers users added value--whether in the form of blog entries, helpful articles, social media posts, or an informative website. When you let possible customers find you, rather than constantly seeking them out, you tend to end up with more qualified leads and earn the credibility and trust you need to establish yourself as the leader in your area.

---

[6] https://blog.rsisecurity.com/the-top-11-rules-of-cyber-hygiene-for-government-agencies/.

[7] Chap. 1 "Cybersecurity awareness" is worth reading in this regard along with the Chap. 2 on "A socio-technical approach to cybersecurity awareness" both show well how agencies can do a great deal before evening considering what type of technological solutions might be used.

This is where customer relationship management, or CRM, software can be invaluable. A CRM can help to:

- **Identify leads**: CRM software tracks actions taken by prospects, such as how often they open your email or visit the website. Use this data to segment prospects in the CRM so you can know which to pursue and which are unlikely to turn into leads.
- **Track your leads**: Your business requires a way to track prospects who have provided contact info. Housing this information in a CRM makes it easy to track who requested to learn more, and who needs some follow up to close the sale.
- **Use automation**: Sending email follow-ups is a good tactic, but without automation, this task can be difficult to manage efficiently. Many systems provide capabilities to send out automatically emails, and to schedule these emails to go out at regular intervals. Obviously, the CRM needs to be organized carefully to send out only emails of interest to customers.
- **Create a unified view of your customers** using CRM software that will serve as a repository for all customer data. This single view of the customer ensures they have a seamless experience when engaging with your company, even if different employees handle each engagement.[8]

7.3.1   Building Credibility and Trust by Creating Comprehensive and Data-Driven Content

The most important issue for the marketing of cybersecurity solutions is the quality of the content being offered. An effective presentation can be more effective than other instruments for promoting more education and awareness of the issues on the part of potential customers. Your priorities should focus on:

- **Improving the "content" your information**: Avoiding old and problematic themes about costs and dangers tends to be counterproductive. Posing different questions and solutions concerning the types of cyber threats one might face over time can be more helpful, particularly to policy officials concerned about their ability to provide services to their communities.
- **Keep in mind the objectives**: To start with you need to figure out your objectives for content marketing and look at why you are doing this at all? What is content going to do for your organization? Will it create more awareness? Can it generate more leads? Would it be useful for improving loyalty and client retention?

---

[8] Comments based on working with a variety of CRM software from open source to Oracle based. One cautionary note is that all these software tools are only as good as the time and resources invested in understanding the overall situation and figuring out what is needed to tracked users. In addition, since they need to be maintained overtime, plans should be made for maintaining and developing the databases.

An important point made by many is that it is not enough to be good at content marketing. Your goal needs to be good at business because of your content marketing.

It is also important to understand the content needs of users—What do clients need to know at the various stages of their journey to buying from you? Segmenting your audience may help define who you have as an audience and what their different needs are.

- **Different blogs** can be useful, for instance, in reaching out and engaging prospective audiences. Whatever activity is chosen should be carefully followed and maintained. Nothing is worse than presenting out-of-date or misleading content to prospective clients
- **Terror tactics**, it is important to remember, they simply don't work in the long run; the effect wears off over time. Giving people real-life situations and questions to frame their thinking around is far more effective in the long run - for example, discussion questions asking potential customers what security solutions does your agency really need?
- **Offering content** whether it's articles, links, or related white papers in download to support your offer is a great way to convince others to look more closely at your offer.
- **Offering content** that can be downloaded is a good way to convert curiosity into substantive leads.
- **Landing pages for content material** (where the material for download can be briefly presented) are useful for users to evaluate whether the material is of interest. These landing pages should also be carefully monitored and, if possible, have someone charged with updating them as necessary. Outdated pages never reflect well on your offerings.
- **Case Studies** are also important for engaging those users who already have a good idea of their problems and what solutions will probably work best. Most such users will appreciate well designed case studies in the local language.
- **Videos** are also an excellent communication tool if planned well. Many executives prefer to watch a video to reading a text. Videos that describe a solution can be the best way to communicate what a cyber security offering does and why it could be useful to your potential customers. As a medium a video may be far more helpful for executives who need more education.
- **Webinars** are also an excellent means to engage with people who are already interested in your solutions. Presenting case studies of your solutions can be an excellent means to allow others to become better acquainted with what you offer as cyber-security solutions.

As you develop your content marketing program you should think of "what sets you apart?" why your solution and not another? There is an enormous amount of content in the market. You should ask yourself will it be useful to your customers. Will others see it as "motivational, inspirational, or otherwise?" What is the core of your content program? Particularly those in policy positions in the public administration will be interested in how what you are proposing could be useful for them.

7.3.2   Metrics of Usage

The saying that "your objectives dictate your metrics" is always good to keep in mind. If you want to produce awareness, try to measure that awareness to give your-self a baseline to measure changes.[9] Four critical metrics are often focused on in this area.

7.3.2.1   Consumption Metrics

Consumption metrics tend to be the easiest to understand. They are the metrics that answer the questions—how many people viewed or downloaded a specific content?

1. Page views are relatively easy to measure with programs like Google Analytics or similar, but you need to take the "page views" in their context. By themselves they are not necessarily that significant in terms of what was achieved and views do not necessarily translate over time into sales.
2. Video views can be seen easily on YouTube Insights or similar programs.
3. Document views: SlideShare, for instance, can facilitate access to data.
4. Downloads can be measured through your CRM or Google Analytics.
5. Social chatter: there are various services available to measure chatter. Here too the context is important to understand the significance of the data. Many agencies (particularly those with out-reach programs) are interested in gauging the reaction of users to their activities.

7.3.2.2   Sharing Metrics

How significant is this content and how often is it shared with others?

1. Measuring how your content is shared impacts two content goals - brand awareness and engagement.
2. Other metrics of quantitative assessment: Often used for assessing, comparing, and tracking performance or production.
3. Likes, shares, tweets: These other tools are often use to make it easy to share data and, depending on the context.
4. Forwarding your email: Google Analytics can provide an indication of where to forward email.
5. Links: These are useful for tracing how people are linked to other web sites.

---

[9] Google analytics is one of the more popular programs to track traffic: what pages are being visited, how much time is spent, what users are returning, etc. To develop an effective search strategy (that is, SEO, search engine optimization) one place to start with would be https://infographic-world.com/ultimate-guide-to-seo/.

7.3.2.3 Lead Generation Metrics[10]

This is one area that, in our experience, tends to be overlooked by sales personnel. Only through a careful monitoring of all eight metrics can one have some idea of what users are doing, who is entering, who enters but leaves, etc. When used these metrics help you answer the critical question "how often does content consumption result in a lead? This includes form completions and downloads, email subscription, blog subscriptions, and the conversion rate. Generally, there are eight important lead generation metrics to measure what users are doing:

1. **Click through rate (CTR)**—This measures the ratio of users who click on a link to the total number of users who view the page, email, or advertisement.[11] It is helpful in analysing how users are reacting, but needs to be considered in combination with other metrics for a full picture.
2. **Conversion rate**—This metric refers to the percentage of users who decide on a certain action.[12] The conversion rate is one of the key metrics to measuring the efficiency of the marketing campaigns.
3. **Time to conversion**—Taken along with the metric of conversion rate these two have an important impact on returning visitors. The better the content being displayed is, the better story you are presenting, the more people will want to return[13] and, thus, reduce the time to a sale.
4. **Cost per lead**—This metric looks at how cost-effective the marketing campaigns are in generating new leads for those in sales.[14]
5. **Leads per channel**—This metric depends on the situation but the "cost per lead" is too general to help understand how to invest efficiently the marketing budget. Calculating the average cost per lead by channel may give you a better idea of how to optimize advertising **costs by concentrating on those channels most likely to provide greater leads to acquisition.**[15]
6. **Month-to-date success**—This metric simply shows where each channel stands against their monthly targets.
7. **Leads to qualified leads**—This metric is used by marketing to measure the quality of the leads they have generated and passed to sales.[16]
8. **Return on investment (ROI)**—The ROI is often used to measure the profitability of an investment. However, at times, it is not the best metric to assess the total value of your campaign.[17]

---

[10] https://www.leadforensics.com/lead-generation-metrics-measurements-for-success/.

[11] Click-Through Rate: Everything You Want to Know and More (cxl.com).

[12] https://dabbling-in-digital.blogspot.com/2021/02/web-metrics-conversion-rate.html.

[13] https://github.com/matomo-org/matomo/issues/6105.

[14] https://simplicable.com/new/cost-per-lead.

[15] https://creassistant.com/how-many-leads-to-expect-per-marketing-channel/.

[16] https://www.klipfolio.com/metrics/marketing/marketing-qualified-leads.

[17] https://sloanreview.mit.edu/article/are-you-using-the-return-on-investment-metric-correctly/.

All these metrics should be analysed together as part of a larger PR campaign to show your users that you understand their issues and needs. Particularly in the public sector it is important to engage with your users and where possible to be involved on a personal level with users. Our experience in Italy is that slick marketing campaigns in the government circles tend to be counterproductive. The metrics described above can lead to a misleading picture of which users are doing what. Evidence should always be balanced with data that comes from personal contacts and user networks.

7.3.2.4   Sales Metrics

Grouped under Sales metrics are a series of quantitative and measurable sales related parameters which you need to monitor in order to understand the effectiveness of your sales campaigns. Tracking these sales metrics is critical for measuring and understanding how best to implement your sales strategy.

Some of the more critical Sales Metrics you could track are:

- What was forecast and what was achieved?
- Your target as compared to the achievement?
- What was the sales growth?
- What possible leads or prospect conversions are there?

As noted previously the average age of employees in the Italian public sector is above 50. Particularly in local government, there remain problems of funds and lack of resources. Employee training continues to be an issue. The availability of recovery funds will certainly alleviate some of these issues, but the most pressing issue of employee engagement remains. In the end, the metrics outline above are indicative but need to be supplemented in the public sector with campaigns that aim at developing direct contact with employees and public officials.

7.3.3   Email Marketing

Email marketing can be an effective way of reaching prospective customers. and if done carefully, of cultivating leads and moving potential users further ahead. An important element, as always, is giving priority to creative and engaging content without exaggerating about what is being offered. Social media tends to be a difficult medium for creating authoritative content. Email allows one to control the message and be more focused. Ultimately, as noted earlier, it's important to improve your content, taking care not to resort to basically "worn-out" themes of impending dangers and doom but to focus on where customers might be in the process and offer them solutions. Awareness fostered through education and persuasion should be the focus of an email marketing campaign for cybersecurity.

You should always try to maintain, in our experience, connections with customers and others with carefully crafted strategies that include simple and direct messages in the national language (in our case, Italian):

- Use well developed statistics, case studies, reports, studies, interviews, and the like that tend to provide readers with a more detailed understanding of important cyber security topics in your emails.
- Compile authoritative whitepapers, if possible, on arguments of interest to clients.
- Ask clients for endorsements.
- Avoid the use of jargon (particularly English based).
- Develop informative content like blogs, downloadable material, webinars, and short video tutorials. Be sure to add links to information about recent attacks and security issues, particularly, where the sources break down complex cyber-security topics for a wider audience.
- Attach summaries of relevant articles from industry sources that help readers better understand what are the hot topics in cyber-security.
- Use leaflets and other material to hand out to people at events or advertisements in print publications (while changing, many in the public sector still use print material).

7.3.4   Try to Educate Your "Bottom-of-the-Funnel"[18] Leads with Interactive Sessions

Webinars can be one of the best ways for those in cyber security marketing to connect with bottom-of-the-funnel leads—those that are fairly far along on the process. A vital part of an effective webinar is the interactive element. It often includes a Q&A session at the end of the presentation that offers attendees the opportunity to ask questions about the topic and the services you are offering. The questions and concerns that users ask during the webinar are also good starting points for developing new content for your target audience. Attendees are already interested in learning more about your solutions and the threats it protects against, and they have usually taken time to do some research. This signifies that they are more likely to be engaged in the topics you are presenting. Therefore, this is an opportunity to advertise other content or encourage demo sign-ups. Even if you decide to pre-record your webinar, you can still accept viewer questions and respond in a follow-up.

---

[18] https://mixpanel.com/blog/bottom-of-the-funnel/. The bottom of the funnel is where you want your leads to be, this is the phase when leads are making the final decisions whether to buy from you or a competitor. One must be very strategic in messaging clients to give them that final push.

If you offer a webinar online, it is always a good to record the content. You can make the recording available later for people who were unable to attend.

If the webinar elicits a good response, then you may also want to use the topics discussed there and create other types of content like blog posts around these topics. In order to promote a webinar and drive attendance, paid channels like LinkedIn and Google "retargeting" ads can be useful but, at times, costly so their use should be carefully analysed.

7.3.5 Up Your Content Strategy Using Paid Campaigns

B2G (Business to Government) campaigns are good for accomplishing two important goals:

- They augment your content marketing efforts
- They also help you direct prospective clients to your demo landing page

In our experience with Italian local governments paid campaigns and inbound marketing are not compatible. If you combine these strategies, you can still manage to create a compelling campaign. For example, as noted in one blog posting from dealsinsight.com, posting studies with some compelling data about a cyber threat can attract interest from interested clients. With this type of asset, time is of the essence—the older the material is, the less likely prospective clients will find it useful. As noted before, updating material is critical when trying to attract the interest of prospective customers. Posting and promoting content through paid channels can let others see results more quickly but still needs to be carefully evaluated in terms of investment versus cost.[19]

One of the major goals of any B2G marketing is persuading prospects to request a demo. While moving prospects to this stage demands a certain amount of work and encouragement, paid campaigns can help accelerate the process for those who are ready to make a buying decision.

Some cyber security companies avoid using paid campaigns due to the competitive nature of paid advertising. A well-developed advertising campaign on Google will probably need a professional to plan, implement, and monitor the campaign. If you do not have a clear idea of what you're doing and what priorities you have, it's easy to spend thousands on cyber security ads and get little or nothing in return. At least within the public sector, networking is crucial for anyone interested in marketing a solution. While public procurement is increasingly regulated within the EU, the competition is quite high and the marketplace is crowded with solutions.

---

[19] https://www.dealsinsight.com/6-powerful-cyber-security-marketing-tactics-that-actually-work/.

7.3.6   Identify the Decision Makers Who Does What?

Identifying and understanding your audience is critical and should be the first step in developing your cyber security marketing strategy. To do this efficiently, many recommend creating "marketing personas". The various metrics mentioned before are useful tools but the key in markets like the Italian is developing personal relationships and networking.

A persona is nothing but a semi-fictional representation of your ideal client. B2G personas not only give a face to your target audience but also provide insights to help you decide which strategies will work best, how to communicate, which marketing channels to use and what kind of messaging will have the desired impact.

If you are marketing to local governments, you need show how what you are offering may help resolve problems facing local governments and their agencies. For instance, compiling blogs which focus on the high-profile cyber security attacks that have hit large accounting and finance firms or Italy's National Health Service leaves local governments thinking they have little to worry about. Like many small-medium sized businesses (SMB) local governments are the weak point in the chain as local officials continue, for understandable reasons, to focus on social and economic programs for their community and pay less attention to cybersecurity issues. Consequently, many attacks are geared at these smaller agencies, often understaffed and ill-prepared.

Creating 2+ personas to cover the different roles that you need to speak to will help develop a targeted campaign. The smaller municipalities will fewer staff. Campaigns usually need to be directed at the larger (medium to large municipalities). You will probably need to target both the IT and the policy sector. In the last analysis, it is always important to research your audience.

7.3.7   Focus on Topics Relevant to the Range of Vertical Services Involved

While "vertical marketing strategies" are commonplace due to their efficacies, many smaller cyber security firms still are reluctant to narrow their focus to a few key vertical elements for fear of alienating a potential prospect that does not fall within those parameters.

Instead, they may focus on the critical areas their solutions address across sectors and treat verticals as an afterthought. This approach does not generally stand out in the B2G cyber security space, at least in the Italian market. Since most buyers of cyber security products work in the same areas, too many vendors end up sounding the same. Buyers of security solutions want "secure end-points, a secure network, the ability to detect a breach, secure software development practices, strong governance, remediation policies in place, and the ability to gain rapid insights when a breach does occur, to name a few". Also, with so many pitching their solutions, the cyber security sector can be noisy and confusing with little distinction in the messaging, which can swiftly become frustrating for buyers.

Sales personnel should always focus on topics their prospects care about, relevant to each target vertical. If a target buyer is a local government agency, better to focus on topics they care about, such as standards and norms in their different areas (health, social services, housing, etc.). In the long run, discussions should always be framed around the buyer's perspective and relevant to their day-to-day operations.[20]

7.4   Closing Comments

How to engage in cybersecurity marketing in the public sector can be difficult to grasp and to effectively employ the available marketing tools. Creating well-crafted content combined with inventive marketing techniques ranging from podcasts, documentaries, cartoons, and videos can be quite effective to attracting and maintaining users in the public sector. Certainly, using these tools requires, at times, significant resources to create a marketing campaign but they can achieve a excellent return. As inSegment points out "increased and higher quality leads generating loyal and long-term customers is just one of the many positives that a well-structured campaign can achieve for your business."

With the growing numbers of businesses in cybersecurity, the industry is becoming increasingly competitive. Those who want to reach prospective customers in the public sector must become more innovative and flexible with their marketing activities. It's your content that matters more than any particular tool or platform.[21] This holds true particularly for government agencies, particularly at the local level where there are generally limited resources and budget. Solutions can be found for staffing and budget issues. The messages conveyed by the content need to be clear. Local officials need informed and authoritative content since they generally do not have the resources or staff to generate it themselves. Typically, local officials have been quite receptive to the approach of CS-AWARE in spending time on developing cybersecurity awareness and subsequently in a collaborative fashion seeking to understand the needs of their communities before even considering what technological issues must be faced.

---

[20] https://www.dealsinsight.com/6-powerful-cyber-security-marketing-tactics-that-actually-work/.

[21] https://www.insegment.com/wp-content/uploads/2017/12/inSegment-5-Successful-IT-Cybersecurity-Software.pdf. The marketing agency inSegment outlined five campaigns that using skilful storytelling made a compelling and effective case for themselves. All the companies behind these campaigns selected in the article by inSegment are "unorthodox and, often, completely new methods to approach complex topics like AI and cybersecurity".

References

Martinez Marias, I., Noupadja, N., & Vander Auwera, P. (2016). *Local and Regional Governments in Europe: Structures and competences.* CCRE CEMR Local & Regional Europe.

OECD. (2009). *Computer Viruses and Other Malicious Software: A Threat to the Internet Economy | READ online.* Oecd-Ilibrary.Org. https://read.oecd-ilibrary.org/science-and-technology/computer-viruses-and-other-malicious-software_9789264056510-en

OECD. (2019). *OECD Economic Surveys: Italy 2019 | READ online.* Oecd-Ilibrary.Org. https://read.oecd-ilibrary.org/economics/oecd-economic-surveys-italy-2019_369ec0f2-en

US Department of Commerce. (2016). Doing Business in Italy: 2016 Country Commercial Guide for U.S. Companies. U.S. & Foreign Commercial Service and U.S. Department of State. https://it.usembassy.gov/wp-content/uploads/sites/67/2016/10/Italy-2016-CCG.pdf

Population of Italy, *ISTAT - National Statistical Bureau,* ISTAT (2022). https://www.istat.it

Chapter 8
Can CS-AWARE be Adapted to the Needs of Different User Groups?

Christian Luidold

8.1 Introduction

Attaining a deeper understanding of a given situation and making a correct assessment requires increasing amounts of data often paired with rising complexities to accurately represent assets and incidents. The analysis of these data volumes generally consumes a considerable amount of time. For fast-paced dynamic environments this can cause considerable concern regarding decision making in a timely manner. Alleviating this obstacle can be complicated by an interdependent environment encompassing different user groups with distinct needs and objectives. In order to facilitate workflow processes affected by existing interdependencies, data can be distributed to different user groups using different means of representation to efficiently meet specific requirements as a form of pre-processing. Providing an additional layer of custom abstractions through interactivity increases the potential for data exploration enabling more individuality while streamlining existing workflow processes. This Chapter presents an approach for exploring the possibilities for adapting the CS-AWARE platform to the needs of different user groups, following an agile rapid prototyping and validation approach in an inter-group setting. The main goal aims at the incorporation of different user groups into a collaborative environment for data exploration, in order to improve decision making (and consequently increase the organization's resilience). For this purpose, a prototype was created based on the CS-AWARE system architecture presented in Chap. 4, which included defining a set of concepts to support the shift in focus. The work described in this chapter extends the contribution of the research conducted in (Luidold, 2021).

The analysis was developed as part of a Master's thesis, and involved expanding the potential application of CS-AWARE to different user groups. The goal of the

C. Luidold (✉)
University of Vienna, Vienna, Austria
e-mail: christian.luidold@univie.ac.at

Master's thesis was to analyse the compatibility between potentially distant fields and roles, within a single shared ecosystem, e.g., combining latest internal cybersecurity events with financial analyses. A simple use case would be the cross evaluation between organizational assets and incidents within a given workflow. For this the evaluation of past events affecting a given asset can be used to infer the urgency of actions based on a set of common identifiers, e.g., type of asset or involved information flows. The main research questions pertaining this work are:

- Collecting information from different sources forms the foundation for effective incident response management. How can big data help improve decision making regarding resilience?
- In an interdependent environment including multiple stakeholders, groups hold different interests regarding a given incident, e.g., information about how to resolve issues for system administrators vs. expected losses for managers. Which data is essential for (re-)presentation to different user groups?
- Different domains may use different visualisation techniques to promote insight generation and awareness, e.g., simple 2D charts vs. hierarchical visualizations. More complex techniques require a higher baseline regarding level of experience in working with visualisations. How should data be visualized and represented to facilitate the needs of different user groups?

The Chapter is organised as follows: In Sect. 8.2, we present a theoretical and conceptual overview of the relevant ideas and the expected benefits of our approach. We combine the evaluation of the requirements analysis conducted in the context of the CS-AWARE workshops and usability tests (Chaps. 2, 3 and 5), with a state-of-the-art analysis of relevant research in the field of visualization for cybersecurity, in order to establish a new set of functional and visual design requirements. The concepts and new requirements are then applied to individual components of the CS-AWARE system architecture. The overview of the additional functionalities is described in Sect. 8.3. Section 8.4 describes the implementation of the main components into a prototype testbed, including the outcomes of a user study carried out to evaluate the functional and design decisions made. The evaluation itself is divided into usability tests followed by a qualitative and a quantitative analysis to facilitate cross-validations with other systems. Section 8.5 provides the conclusion of the chapter, including a summarization of the findings, as well as a discussion of the results.

8.2   Concepts and Requirements

In order to investigate the research questions a list of prerequisites was drawn up that would allow an extensive study to be carried out. This list consists of a set of supporting concepts as a form of framework conditions derived from the research questions, as well as functional and visual requirements. The requirements analysis was conducted in part of by using the data gathered during the workshops and usability tests in the CS-AWARE project (detailed in Chaps. 2 and 5), involving domain experts working in local public administrations (LPAs). Additionally, a short interview was held with domain experts working in the field of cybersecurity. Lastly the results of a state-of-the-art analysis of latest relevant publications regarding major gaps in functionality of cybersecurity tools and visualization techniques (Luidold & Schaberreiter, 2020) were incorporated to establish a baseline on which to expand upon.

8.2.1   Supporting Concepts

Derived from the research questions four major concepts were identified to support the definition of requirements for incorporation of different fields and user roles in the CS-AWARE system. These consist of:

- Collaboration & Communication, for conveying information primarily in an intra-group setting. The goal of the concept in the context of this approach focuses on more visible interaction possibilities within an organization.
- Multi-Stakeholder Involvement, for increasing the engagement in an inter-group setting in order to facilitate communication by addressing situations and incidents from different angles.
- Multi-Stakeholder Visualization, for meeting the interests of different user groups, while providing the same data basis. The goal is an increased consideration of different requirements regarding potentially diverse user groups.
- Situational Awareness, as the perception and correct assessment of a given situation. The goal hereby lies in improving the processing of information content from a variety of sources to facilitate decision making.

8.2.1.1   Collaboration & Communication

The concepts of collaboration & communication represent the system's ability of facilitating collaborative actions, as well as intra-group and inter-group communication. By incorporating additional functionalities, e.g., data sharing, these concepts lay the groundwork for effective security.

The connection between individuals of the same group (e.g., security analysts) within an organization through additional communicative channels hold the potential of supporting internal workflow processes and collaboration. This type of communication involves a shift from physical in-person communication to online communication, while still providing access to the whole knowledge base of the organization. The main approach for achieving this objective is to incorporate tools for including subsets of data through the communication channels, by providing either static snapshots or dynamic references to given events or analyses.

8.2.1.2 Multi-Stakeholder Involvement

Multi-stakeholder involvement aims at facilitating processes requiring the interaction with individuals, groups, organizations, or authorities about past or current incidents. It can be divided into two main categories:

- Internal stakeholders, e.g., various departments within the organization, and
- External stakeholders, e.g., clients or partners using the goods and services of the organization.

Concerning the first type, the goal regarding the involvement of internal stakeholders can be interpreted as how a given user group (e.g., system administrators) can communicate different specific types of knowledge (e.g., needs or incident reports) to other selected user groups (e.g., accounting, or public relations). A use case example is the communication of inhibited services affecting business continuity caused by an incident to other departments mostly by notifications. Depending on the organizational structure this type of involvement can be used to foster future communication & collaboration.

The second type focuses on how to effectively communicate knowledge to external stakeholders (e.g., clients or supervisory authorities). Through the course of an organization's operation the occurrence of significant events can be expected. Through the last years a growing number of National States have implemented new legislations to oblige organizations to notify other organisations and national authorities about events, such as personal data breaches, in a timely manner. The General Data Protection Regulation (GDPR) is a prime example of a shift towards a more consumer protection-oriented environment. In order to alleviate the process of communicating incidents, the required data has to be pre-processed to account for the distinct needs and interests of the respective stakeholders (e.g., supervisory authorities, clients) including only necessary subsets for each group.

8.2.1.3 Multi-Stakeholder Visualization

Multi-stakeholder visualization aims at providing different views of a given state to distinct user groups using the same underlying data. A general example can be given through the provision of data consisting of organizational assets to multiple distinct user groups (e.g., system administration or accounting). Similar to the concept of multi-stakeholder involvement the visualizations of the data source need to take the needs and interests of each involved user group into account. While the system administrators might focus on the health of individual assets, the accountants might aim their attention on the documentation of the financial value of assets. In this regard being able to translate different aspects of data for multiple user groups facilitates communication and generates a more in-depth understanding of existing connections between points of interest.

While visualizations are useful for creating an overview of potentially extensive amounts of data, major attention should be drawn to the usability and experience regarding individual visualization techniques in combination with data types. An example would be the usage of hierarchical data and subsequently hierarchical visualizations (e.g., tree maps or sunburst charts), the latter of which can seem unintuitive and cumbersome for users not accustomed to these types.

8.2.1.4   Situational Awareness

Increased situational awareness in the context of cybersecurity aims at raising the perception and judgment regarding activities and vulnerabilities. This can generally be achieved by incorporating information from various sources, both internal, as well as external. This also encompasses the aspect of collaboration outside and inside of communities sharing knowledge from latest experiences or analyses (including reports or malware samples).

8.2.2   Functional Requirements

The functional requirements were gathered in part from the CS-AWARE workshops as described in Chap. 2 with LPAs as well as by conducting a state-of-the-art analysis from included evaluation cycles and expert interviews. The included user groups consist of:

- Data analysts: This user group is defined by its need to explore and analyse a potentially significant amount of data as provided by internal logs or connected partners and communities. The data analyst group is derived from the CS-AWARE system administrator group with a focus on analysing event data.
- Managers: The management group is characterized by a focus on financial aspects of the current situation, as well as evaluating additional investments in a cost vs. impact manner.
- System administrators: The attention of system administrators is focused on low level business continuity regarding the availability of existing organizational assets. Similarly to the data analysts, the system administrator group is derived from the CS-AWARE system administrator group with a focus on monitoring internal assets.

The requirements stemming from the workshops with individual partners working in LPAs (see Chap. 5) was mapped directly to the requirements of a paradigmatic prototype. The mapping focused on internal artefacts, behaviour, and objectives, which could be evaluated in a closed environment, therefore excluding external groups (e.g., external service users). The major functional requirements selected were:

- A *monitoring system* displaying the current state (e.g., selected KPIs) within the organization and current activities outside the organization carrying the potential of affecting the organization's systems and services.
- A *reporting system* generating automatic periodic reports including latest snapshots in the form of statistical analyses (e.g., number of attacks, occurrences by defined categories, current trends, affected services).

Regarding the state-of-the-art analysis, research incorporating organizational and governmental needs was evaluated (Luidold & Schaberreiter, 2020). The examination of published work was divided into the investigation of functional and design decision making (e.g., types of charts, interactive features), and the feedback acquired from the evaluation methods that these studies applied (e.g., usability, request for additional features). Another aspect was the analysis of user stories (Chap. 3) and other interviews with domain experts, which created additional insight into common processes as well as exceptional and unique events.

8.2.3   Visual Design Requirements

Taking into account the different user groups needing to collaborate during given incidents results into a vast landscape of different needs to be accommodated for. The possibilities of provisioning the means to fulfil those needs are just as plentiful. In order to create a baseline regarding data visualization and data exploration, the analysis of functional requirements, as described earlier in Sect. 8.2.2, was extended to include visualization techniques (e.g., simple two-dimensional charts, geometrically transformed displays) as described by Keim (2002), chart types (e.g., pie charts, bar charts), as well as interaction possibilities (e.g., brushing & linking, zooming & panning).

Dividing the visualization landscape into the corresponding user groups leaves us with a set of both shared, as well as unique design decisions. Starting with the shared principles the most common desire among the evaluations was the addition of a structured textual representation (e.g., within a data table) of a given event. Where included in the prototype, this visual functionality earned heavy usage from domain experts by providing a complete depiction of the selected data entry. In case where this functionality was not present, the participants had to rely on either combing the log files as a manual task or accept the potentially limited insight of visualizations, as those might not deliver the full picture of the data. An example would be the depiction of a lengthy entry within a log file, which might not be able to provide included descriptions, or unique values (e.g., identifiers, hash codes) within vast amounts of data.

Another shared design decision is the focus on simple chart types. While each individual user group might have one or the other preferred visualization with a higher degree of complexity, a translation to other user groups might not be possible. Furthermore, those individual preferences might be limited to just a single member of a group, therefore inhibiting the readability for other group members. The simplest chart types consist of bar charts, line charts, pie charts, and scatter-plots. While the resulting visualizations using those simple types are generally easy to comprehend, they might be prone of falsifying the created insight by either excluding potentially important properties (e.g., by not supporting the necessary number of dimensions) or by merely providing arbitrary values (e.g., by depicting relations or hard to read values instead precise values) among other things.

Comparing the unique design decisions between the defined user groups facilitates the recognition of distinct needs. Starting with the data analysts their requirements focus on investigating activities and vulnerabilities using various resources including methods for cross-evaluation. In terms of visualizations, chart types taking the dimension of time into account (e.g., bar charts) are regarded as indispensable when analysing relationships and historic actions. In order to gain a quick overview over a given network system administrators rely on a node-link diagram to assess assets in terms of health and interdependencies to other assets. Participants from the economic sector stated a general focus on simple intuitive charts (e.g., pie charts). Although, contrary to exploring data for in-depth analyses, employees generally tend to use charts for the visualisation of results, which are then exported in form of reports or statistics. With this in mind employees from the economic sector tend to be more accustomed to working with raw data or data tables in order to obtain results and sufficient situational awareness.

8.3 System Architecture

This section describes the prototypical architecture of a conceptual testbed used for exploring and evaluating the supporting concepts as well as the functional requirements. The individual components presented build on the components defined in Chap. 4 .2 with a set of alterations done in order to accommodate the concepts specified in the previous section.

8.3.1 System Overview

The paradigmatic prototype makes use of the CS-AWARE framework as presented in Fig. 4.1. A similar high-level view describing the individual components and information flows can be seen in Fig. 8.1. The grey components are defined by the CS-AWARE framework, whereas the three coloured components (Data Processing, Explorative Data Visualization, Communication & Collaboration) describe additions and adaptions conducted. The information flows were adapted correspondingly to the new ecosystem. The distinct layers are unmodified from the CS-AWARE framework.

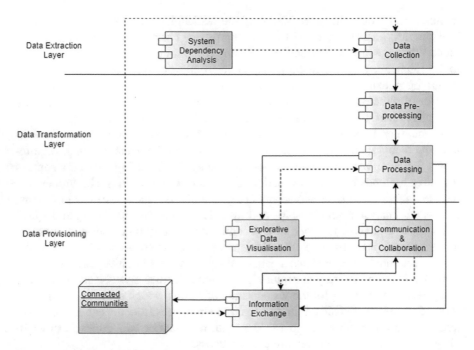

Fig. 8.1 Components and information flows of the prototype testbed

8.3.2 Data Collection and Information Sharing

The data collection process in the context of the paradigmatic prototype is divided into internal data as well as external data sources. The latter is used by incorporating data feeds from the Malware Information Sharing Platform (MISP)-project (C. Wagner et al., 2016). MISP is a community-driven open-source project for sharing data of ongoing events (e.g., reports, activities) concerning malware or threat actors. It implements a five-tier hierarchy regarding the publication of information permitting an organization internal-only usage of corresponding internal data. Individual additions can be shared with other selected community members or made public.

In the context of the paradigmatic prototype the internal data is provided by the given network structure adapted from conducted CS-AWARE workshops. It consists of a network layout of a local public administration (LPA) adapted for privacy purposes and expanded by supplementary fields to facilitate data processing and extend the prototypical implementation by additional functionalities. For simulation purposes the added fields consist of:

- x_status: A custom property depicting the given state of the corresponding asset in terms of health or availability, similarly to the "service status" field in regard to the Windows Service Status Enumeration.

- x_history: A custom object consisting of historical data of the corresponding asset cross-referencing events either from logfiles or from events provided by the MISP feed.

A semi-synthetic set of financial data was created according to Gonzalez-Granadillo et al. (2016, 2018) on how to compare the financial and operational impact of defined mitigation actions to facilitate decision making. In the context of the paradigmatic prototype data regarding the network layout is incorporated into the framework. For data exploration purposes the resulting data was unified into a homogeneous data structure.

8.3.3 Data Processing

An organization generally logs a significant amount of data each day with automated methods in place to exclude ordinary entries and summarize data for facilitating the creation of an overview over a given state. While these routines prove useful for focusing on relevant data, those routine might not be exactly fool proof and could potentially inhibit analyses by human users. The functionality to be able to interact with the underlying data provides additional possibilities to support insight creation. These interactions can be realized by simply adding parameters to tasks for further customization. An example would be the use case regarding the automatic report generation. For periodically planned reports (e.g., weekly, monthly) a standard set of parameters is enough for evaluating a given snapshot and facilitates the comparison to historical data. Including the functionality to impose a specific focus on the analysis provides added value for individual user groups with instances being the incorporation of only a subset of the data consisting of the most important events or excluding protected data for the result to be shared with other internal groups or external communities and organizations. Especially the latter example lowers the barrier of communication and information sharing between dependent stakeholders.

Another use case constitutes on-demand data exploration. This is depicted by the prototype's ability to process semi-custom queries from users to enable individual data analysis tasks either by interacting with the prototype directly or by other connected systems through the provided API. Examples of connected systems can be information sharing platforms (e.g., MISP) or communication applications (e.g., Mattermost 2015). The semi-custom queries are defined by providing a template of possibilities either by brushing & linking within the prototype's visual interface, or by customized text-based commands passed to the API.

8.3.4 Explorative Data Visualization

When it comes to analysing the ongoing state within an organization, the current approaches record a shift from loading snapshots (e.g., logfiles) as a form of static mapping to real-time detection using dynamic mapping as described by Aigner et al. (2011) and adopted by Wagner et al. (2015). Hereby the terms "static mapping vs. dynamic mapping" indicates whether a user has to interact with the system to be shown the latest data. This can be imagined as having to click a button in order to refresh the views in terms of a static approach vs. automatic data streaming regarding a dynamic manner in which case it is to be expected from the rendered visualizations to be rebuild by the system. While the latter is often more valuable, as it truly incorporates events upon their logging, this results in performance issues in the instance of high traffic. Accepting a trade-off between a continuous livestream and preventing a bottleneck on end user devices results in a near real-time visual analysis.

A dashboard filled with charts and tables depicting different aspects of the underlying data is a good start to create an overview over a given state. It becomes even more valuable if the individual visualizations can be interacted with. By incorporating the concepts of zooming & panning, as well as brushing & linking this higher degree of interactivity can easily be achieved. As the term implies, zooming & panning provides straightforward means of interacting with the charts (e.g., to enlarge parts of a visualization), the latter term provides a visual filtering alternative translating to the underlying data. Brushing & linking connects the individual charts and turning those into filters in order to dynamically impose additional conditions on the underlying data resulting in the creation of subsets. These selected changes force the connected charts to be redrawn enabling simple visual data exploration.

Approaches in terms of number of charts, as well as types of charts are analysed from a research perspective & usability perspective pointing out the discrepancies between them, e.g., people from the financial sector are used to more common graphs like pie charts with scatter plots as the most complex type, whereas data analysts might find additional insights in hierarchical visualizations (e.g., sunburst charts). Despite the possible limitations regarding individual experience and preferences, the prototype used various approaches in terms of visualizing charts paired with a varying number of charts in each view, as well as interactive features. The goal was to assess the interactions by the participants during the evaluation phase as well as their feedback. Core points of the evaluation were the usability of the prototype testbed, the readability of the views, and the individual feedbacks concerning the comparison of the prototype testbed to the tools and ecosystems in their own work environment.

8.3.5 Communication & Collaboration

This component embodies an extension to the regular communication & collaboration environment within a given group, organization, or community. It facilitates workflows by incorporating means to interact with collected event data in a similar explorative way to the data visualization component. The interaction possibilities encompass single event interactions and event summaries. Those interactions are based on creating static snapshots of the data for accurate referencing and historic relevance.

The single event interactions are defined as providing detailed information of a selected event by either highlighting selected characteristics of the event itself or presenting knowledge of similar events. The latter can be achieved by the prototype's analysis of included metadata (e.g., tags). This enables users to efficiently search for related cases by examining other events regarding their behaviour (e.g., effect on organizational assets) or historic actions (e.g., mitigation actions conducted). The results aim at facilitating decision making in order to increase the organizations resilience.

Event summary interactions, as the opposite of single event interactions, aim at providing an overview over a given state within the organization (and/ or outside, if applicable). This enables users to group data into smaller subsets used for facilitating stakeholder-oriented report generation or for further data exploration tasks. In the latter case the results can be used for single event interactions (e.g., for outlier analysis).

8.4 Prototype & Evaluation

This Section describes the implementation and the evaluation of the prototype testbed. Hereby the prototype system refers to a software prototype encompassing only the individual views or dashboards and focuses only on the aspect of visualization. The prototype testbed refers to the entire environment, particularly the prototype system and its connected service components, e.g., data provisioning and communication. The component of data collection and information exchange is supported by the usage of a MISP (Malware Information Sharing Platform) instance as means to exchange knowledge between organizations and communities about analyses regarding malware, behaviour, and latest activities. The component of communication and collaboration is supported by the inclusion of a Mattermost instance into the prototype testbed containing a set of custom bots and slash commands to execute automatic tasks in order to facilitate data exploration in a static messaging environment.

8.4.1 Implementation

Regarding the implementation of a testbed in order to evaluate the concepts analysed, the developed prototype was created using Node.js running in a virtual Docker environment similarly to the CS-AWARE implementation. The connected services for alleviated data sharing and communication were provided by MISP and Mattermost respectively, each running in their own virtual environments. The communication between the Docker containers and takes place via a defined REST API.

8.4.1.1 Prototype System Overview

The prototype system consists of three individual views for each addressed user group encompassing the asset view from a system administration perspective, the event view for visual exploration of event data, and the finance view for correlating the financial worth of assets and actions. Each view was built with a set of differences ranging from subtle to more visible changes.

The asset view, as shown in Fig. 8.2, is mostly composed of simple to read chart types, i.e., bar charts and row charts, combined with a data table. Additionally, a node-link diagram depicting the organizational network is used to provide an overview over the current state. Including the data table, the asset view incorporates a total of five charts.

The event view, as shown in Fig. 8.3, is entirely composed of simple chart types, i.e., bar charts and pie charts, with the addition of the data table. Together with the different colour scale, the significant visual changes to the asset view, as presented in Fig. 8.2, are the added legend to the pie charts, which are linked to their corresponding pie slices. The bar charts depict the same dimensions and act as a

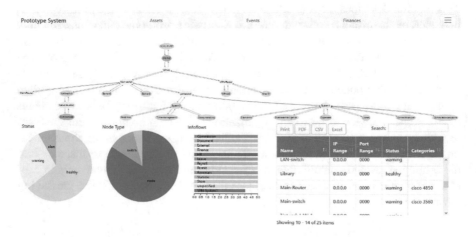

Fig. 8.2 The asset view consists of a node-link diagram depicting a given network in the center part as well as a set of simple interactive charts and a data table in the bottom part

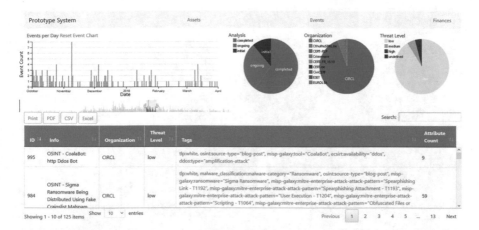

Fig. 8.3 The event view consists of two bar charts (the upper focus chart for more granular exploration and the lower overview chart as a form of mini-map) and additional pie charts with the data table at the bottom

combined entity consisting of the upper (focus) bar chart enabling zooming & panning, and the lower (overview) bar chart always displaying the entire range of events (greying out all events outside the selection) to alleviate navigation. Furthermore, the data table of the event view implements pagination, as opposed to infinite scrolling from the asset view. Including the data table, the event view incorporates a total of six charts.

The finance view, as shown in Fig. 8.4, holds the most diverse types of charts consisting of a pie chart, a sunburst chart, a row chart, a bar chart, and three scatter plots with the data table at the bottom. Furthermore, the sunburst chart is the most complex chart type regarding intuitiveness used throughout the entire prototype system. The finance view is characterized by combining different features from the other two views, primarily through additional labelling regarding the axes and through having the pie chart possessing its own legend in comparison to the sunburst chart, which only relies on its labels for their corresponding segments. As the finance view is holding a total of eight charts (including the data table) it is the only view, which does not fit onto a regular screen without the need of scrolling.

The underlying data for the prototype testbed was of peripheral importance for the analysis of the supported concepts and the defined requirements. Knowledge of the individual domains was welcomed, but not a necessity due to the different domains presented, as well as participants being employed in different sectors. The individual data sources consist of:

- The asset network provided by CS-AWARE adapted for the prototype testbed. This data was used primarily for the asset view and in part for the finance view.
- Event data from live feeds provided by the MISP project used for the event view.
- Synthetic data created according to Gonzalez-Granadillo et al. (2016, 2018) and adapted for the prototype testbed used for the finance view.

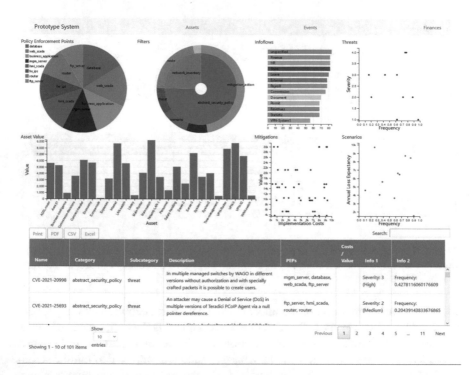

Fig. 8.4 The finance view consists of one pie chart, one sunburst chart, one row chart, one bar chart, and three scatter plots, with the data table at the bottom

The included data sources were used across the provided views in order to facilitate cross-referencing individual data entries. This supports analysing the effective range of occurrences and potential disruptions.

8.4.1.2 Design

Regarding the visual design decisions, the individual views were built with rising complexity in order to assess the willingness to work under different criteria, focusing on:

- The number of charts per view including related aspects concerning the end user's enthusiasm about the entire dashboard to fit on the screen without scrolling.
- The type of charts used to visualize the underlying data. The main aspect here lies in analysing the acceptance of more complex charts escaping the boundary of the most common charts (bar charts, line charts, pie charts, and scatterplots).
- The degree of information provided by individual charts. This can be achieved by choices regarding the inclusion of additional characteristics legends or more descriptive labels and tooltips.

The visualizations themselves were created using D3.js for the creation of visual documents mainly in combination with crossfilter.js used for the exploration of multivariate datasets through interactive means (e.g., clicking, brushing & linking) and DC.js acting as their wrapper.

8.4.1.3   Functionality

The prototype system's views all incorporate the functionality of brushing & linking to facilitate data exploration. Figure 8.5 depicts an example by having the different chart types act as interactive filters in a dynamic environment. Furthermore, the node-link diagram in the asset view, similar to the bar charts in the event view support zooming & panning to enlarge the visualizations in order to increase readability. The search functionality of the connected data table provides an additional sub-filter for querying the individual searchable fields (depicted or hidden) of the underlying data. The results can be exported as a list either as a regular text document, or in the form of a PDF file, CSV file, or Excel file.

Selecting multiple filters from different charts can visually alter the displayed data significantly, as each chart presents the same shared subset of data. This results in either the exclusion of entries by charts without active filters or by greying out parts of charts currently applying filters (as depicted by the pie charts and the row chart respectively, seen in Fig. 8.5).

In order to provide additional information of individual data entries, tooltips can be implemented by hovering the cursor over either a node or an edge (or a given row

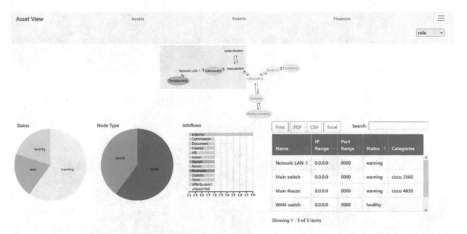

Fig. 8.5 Brushing & linking presented by first selecting the categories "External", "Payroll", and "Revenues" in the row chart, which causes the other charts to rebuild while ignoring (greying out) unrelated data entries. Afterwards Brushing over the node-link diagram afterwards applies further filters on the underlying data. Depending on the organizational asset structure different layout engines (here cola) can be selected from the drop-down menu accessed by clicking on the menu on the top right corner

within the data table), as exemplified in Fig. 8.6. Regarding of the pragmatic prototype system the tooltips are only implemented in the context of the node-link
diagram.

Providing a more structured form of detailed information about an individual
data record, the addition of a model was chosen due to more extensive possibilities
for data exploration and insight generation. In the context of the pragmatic prototype system's asset view, the modal consists of related information regarding the
selected asset and an interactive data table of related events linking to the entry
within the community-driven platform (MISP), as depicted in Fig. 8.7. A button
redirecting the user to a corresponding Mattermost channel allows users to inspect
historic information and decisions, as well as share questions and insights.

In order to facilitate intra-group communication and decision making, a self-
hosted Mattermost instance was chosen to be included into the prototype testbed.
Additional configurations of slash commands extended by custom bots were used to
alleviate obstacles regarding usability and interaction processes with the data itself.
Furthermore, Mattermost provides a mobile application, which facilitates communication and collaboration from different devices.

Considering the CS-AWARE requirements analyses (presented in Chap. 2)
called for the inclusion of monitoring and reporting systems ranging from generating periodic reports to notifying system administrators and service users about
events affecting internal services, the prototype testbed uses a customized
Mattermost instance to meet those requirements. Specifically, by using the API of
the prototype system to explore and summarize data, the results can be formatted
and send to the corresponding channels with minimal user interaction, as exemplified in Fig. 8.8. Supported functionalities include:

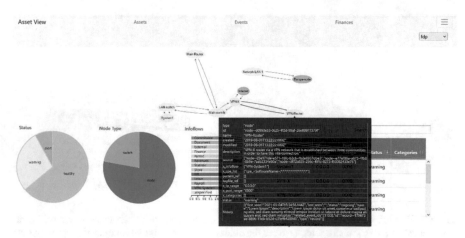

Fig. 8.6 Hovering the cursor over a node or an edge provides a table of information within a
tooltip. Here the layout was changed to "fdp". The selected filters (slices) from the left-most pie
chart cause the node-link diagram to build multiple sub-graphs due to a lack of existing connections between individual nodes

Fig. 8.7 By double-clicking on an individual data record (in the asset view either through the node-link diagram or the data table) additional information can be displayed through a modal. In case of the asset view the related information can be accessed through hyperlinks (events affecting the asset) redirecting to a connected MISP instance. The "Open Chat" button connects the user to the corresponding channel of a Mattermost instance

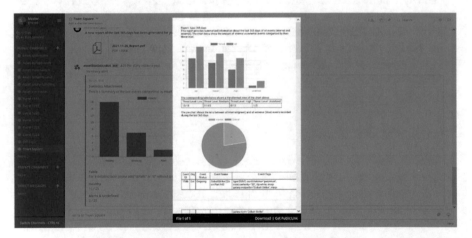

Fig. 8.8 Using custom slash commands request the server can be tasked to provide static information either as regular formatted messages from different bots or exported as PDF

- Report generation in user defined intervals tailored to the needs of different user groups by using either complete sets of data or by creating and summarizing subsets (e.g., by time frame, severity, etc.). The resulting reports can be distributed through various means, e.g., by email or through Mattermost channels.
- Static data exploration, which supports custom user queries to create a snapshot of the specified results. The main difference to the three views created in the context of the prototype system is the lack of dynamic exploration features (e.g.,

dynamic mapping) and techniques (e.g., brushing & linking). Results can be formatted in various fields, tables as well as attachments (e.g., charts).
- Data manipulation in a simplified form of changing writable attributes. This includes overwriting descriptions or threat levels (if necessary), adding or extending comments and entries related to an event or asset, as well as proposing solutions to given circumstances.

Similarly to Franklin et al. (2017), the inclusion of a messaging application provides the support of adding custom notes, screenshots, as well as additional formats through plugins.

8.4.2 Evaluation

For the evaluation phase a user study consisting of ten participants from different fields with varying degrees of experience was conducted. In order to gain distinct qualitative results, the participants were divided into three subgroups to gain deeper insights regarding their individual usage and requirements.

The user study itself was divided into the practical usability test by providing an interactive playground of the testbed described in Sect. 8.4.1, paired with a set of tasks to solve, a qualitative analysis in the form of a semi-structured interview, and a quantitative analysis in the form of a questionnaire. Particular focus was aimed at the intuitiveness of the views and the added custom functionality of the connected services.

8.4.2.1 Participants

In order to enable an effective comparison of the results from the evaluation phase, the participants were divided into three subgroups according to their field of work. As the initial design of the system architecture focuses on increasing the resilience of a given organization and facilitating decision making, a major focus lied on the participants working in the field of cybersecurity research resulting in them constituting the focus group but was later extended to encompass a more diverse set of participants. The individual groups are described below:

- The focus group (FG), consisting of three cybersecurity researchers (having less than 3 years, 3–5 years, and 10+ years of experience respectively),
- the IT group (IT), consisting of the three cybersecurity researchers from the focus group above and two employees from the IT sector (having 3–5 years of experience) for an additional layer of abstraction,
- the economy group (EG), consisting five employees (with four participants having 3–5 years and one having 10+ years of experience) working in banking, marketing, and tax consultancy, and
- the whole group (All), consisting of all (ten) participants.

In the context of the evaluation phase the above user groups are ranked according to their relation to the, with the analysis primarily being tailored to the focus group. The number of participants reflects a suitable sample size. Given the proximity between the focus group and the IT group, the defined user groups display a uniform distribution.

8.4.2.2   UI/UX Test

In order to validate the design decisions and theories made, the participants were asked to interact with the prototype testbed before continuing with the interview. During the entire process the participants were requested to think aloud about what they wanted to achieve through their interactions and to compare what type of result they would expect with the result they received. The tests were divided into three parts with a combined duration of on average 30 minutes.

The first part was the initial first reaction and unsupervised discovery of the prototype testbed. Participants were not given any kind of explanation on how potential functionalities were included and how to use it in order to determine intuitiveness and the expected usability for first-time users, as well as ascertain additional expected and desired functionality. The participants could navigate through the individual views and the connected services as well as interact with the underlying data in a preliminary exploration, although during this part unclear aspects were noted to be answered during the next step.

The second part was divided into two subparts starting with the introduction into the provided functionalities and the navigation between all included components as depicted in Table 8.1. Afterwards the participants received a set of short tasks to solve categorized into:

- The exploration of the data as well as the generation of subsets for further analyses, e.g., "Find all events which occurred from two months ago affecting organizational assets related to internal finance processes".

Table 8.1 Main features of the usability test regarding the prototype testbed

Summary	Usability study, all groups
Use cases	(1) interactive data exploration through the individual views using brushing & linking. Task: Create a subset including only data records matching given criteria (e.g., infoflows affected by vulnerable assets, internal high priority incidents) (2) incident handling using static data exploration through Mattermost. Task: Find a solution to an incoming incident, by investigating similar (past or active) incidents, including knowledge shared by connected communities
Components	Communication & Collaboration, data processing, explorative data visualisation
Participants	(3) cybersecurity employees, 2 employees from the IT sector, 5 employees from the economic sector
Method	Cognitive walkthrough
Outcomes	Insights regarding intuitiveness, usability, and readability

- The handling of new events by correlating them with other (past or current) organizational events and/ or events shared by the connected communities before solving them, e.g., "Provide a solution for event #1234 affecting the organizational assets ABC and DEF (including their corresponding infoflows) by searching for related information".

The third part consisted of a semi-structured interview about the overall impressions of the prototype testbed pertaining to the individual views, the included functionality, and the usage of the connected services. Furthermore, the participants were asked to compare the prototype testbed to similar software solutions used in their work environment on a more abstract level regarding its usage potential. The results of the UI / UX tests are summarized in the subsequent Sect. 8.4.2.3 including notable remarks by the participants.

8.4.2.3   Qualitative Analysis

The qualitative analysis was conducted by analysing the interactions and verbal statements of the participants during the user study. The results of the evaluation presented constructive feedback regarding a set of visual and functional design changes focused on:

- Additional descriptions to the individual charts regarding the displayed dimensions and results.
- Simplification of slash commands regarding data exploration.
- Additional information regarding individual data points or bins within charts.

In terms of design decisions, the inclusion of labels for the individual charts was regarded as a core requirement regardless the simplicity of the chart or the depicted dimensions. Furthermore, an additional description for each visualization was desired describing the visualized dimensions (in the context of the evaluation, knowledge of the underlying data was not required). These were more desired for scatterplots and bar charts, although less important for pie charts. The node-link diagram specifically in combination with the included tooltips was regarded as sufficiently precise without the need of additional descriptions.

Noteworthy statements pertaining constructive criticism focused particularly on the types of charts used. The focus group and the IT group felt confident in the interpretation of the visualizations as well as the assessment regarding their values. As an exception, one participant from the IT group voiced a negative opinion on the usage of pie charts. Despite their intuitiveness and simplicity, pie charts individually hold the potential to inhibit a correct judgement, as they might distort the represented picture through the lack of relation to other aspects of the data. The economy group on the other hand preferred the usage of simple graphs (e.g., bar chart and pie charts), but every member of this group voiced concern or trouble with the interpretation of the sunburst chart by either themselves and/ or colleagues working in their field, whereas members from the other groups praised its inclusion as reliable example for hierarchical charts. Particular statements concerning the sunburst chart from the economy group were "I don't understand what I'm looking at.", and "It's okay-ish for me, but I couldn't possibly include a chart like this in a report for my boss." This disparity regarding individual intuitiveness conforms to the question whether to design a system adapted for multiple differing user groups or focused on a set of closely related user groups. The consensus regarding the prototype testbed lies in a clear distinction of individual views as to avoid the loss of functionality and detailed visualizations even at the cost of interoperable analysis.

In terms of interactivity all groups were positively impressed by the inclusion of brushing & linking to create subsets of data. Particularly participants from the economy group were stated this technique to be especially useful for outlier detection and analysis as their work environment generally included handling larger amounts of raw data or data tables.

8.4.2.4   Quantitative Analysis

For the quantitative analysis the participants were asked to fill out a questionnaire pertaining their experience with the prototype testbed, as well as a combination of abstract work-related experiences (e.g., which types of visualizations are employed?), preferences (e.g., which types of visualizations are preferred?) and a set of demographic information.

The assessment of the system usability scale (SUS) according to the formula described by Brooke (1995) was applied for each individual view. In order to confirm the SUS score the participants were additionally asked to rate the views on a Likert scale from 1 (worst imaginable) to 7 (best imaginable) as proposed by Bangor et al. (2008) as a form of extending the assessment. The results for each view according to the individual user groups are summarized in Table 8.2.

The results concerning the achieved SUS score are characterized by a preference for cleaner views by all participants, even though the navigation and interactivity might come as unintuitive and would require a brief introduction. As in the case of

Table 8.2 Individual results of the quantitative analysis by user group and views. The average usability score on a Likert scale from 1 ("worst imaginable") to 7 ("best imaginable") is provided for correlation (correlation coefficient provided as $r = 0.806$) to verify the calculated SUS score

	Asset view	Event view	Finance view
Calculated SUS score (FG)	80.00	73.33	64.16
Calculated SUS score (IG)	82.50	79.00	67.00
Calculated SUS score (EG)	70.00	60.50	42.00
Calculated SUS score (All)	76.25	69.75	54.50
Average usability score (FG)	5.67	5.33	5.33
Average usability score (IG)	5.40	5.40	4.80
Average usability score (EG)	5.40	4.80	3.60
Average usability score (All)	5.40	5.10	4.20

the asset view participants praised the automatic adaptation of the node-link diagram through brushing & linking with the other charts despite initially not realizing the additional functionality pertaining the node-link diagram itself, i.e., zooming & panning for enhanced navigation, and change of the graph layout to address the problem of unfitting layouts as stated by McKenna et al. (2015).

Bangor et al. (2008) conducted an empirical evaluation of SUS results, which can be used as a benchmark for comparing the results achieved of the prototype testbed, specifically the individual views. More precisely, the prototype testbed is compared against the category of internet-based Web pages and applications with the latter providing a mean (M) score of 68.05 with a standard deviation (SD) of 21.56 of a total of 1180 analysed surveys. Juxtaposing those results with the calculated SUS scores from Table 8.2 shows a favourable tendency for all user groups regarding the asset view (M_{FG} = 80.00, third quartile) and the event view (M_{FG} = 73.33, second quartile), while the finance view (M_{FG} = 64.16, first quartile) was, as anticipated according to the design decisions described in Sect. 8.4.1, less well received, although still managed to stay in a reasonable range.

The effects regarding the remaining supporting concepts, as described in Sect. 8.2.1, on the participants a small set of changes in definition was applied to facilitate the evaluation. Specifically, the supporting concept of communication & collaboration was divided into its two respective parts in order to provide a higher degree of freedom of expression on a lower level without affecting the reception of concerning the functionality of other components. Additionally, due to the interactive features included with the visualizations the analysis of the concept of multi-stakeholder visualization focused on data exploration according to the individual perception by the participants. The individual results categorized by user group of those perceptions are depicted in Table 8.3.

The results as seen in Table 8.3 regarding the perceived effects of the concepts were measured on a scale from 0 ("very unhelpful") to 4 ("very helpful") well received. Specifically, data exploration reached the best mean grade (M_{All} = 3.50), which was also the most praised aspect during the UI / UX tests, whereas the aspect of collaboration received comparatively lowest score of slightly more than just helpful. The highest individual score received the concept of situational awareness by the focus group (M_{FG} = 4.00) due to the dynamic mapping approach as described by Aigner et al. (2011)

While translating the results of the individual view into a ranking were to be expected due to the explorative design decisions made, the results of Table 8.3 depict the prototype testbed to have unique strengths for each user group with

Table 8.3 Overall impact on the analysed concepts perceived by user group ranging from 0 (very un-helpful) to 4 (very helpful)

Group Concept	Focus Group	IT Group	Economy Group	All
Collaboration	3.00	3.00	3.40	3.20
Communication	3.33	3.60	3.20	3.40
Data exploration	3.33	3.60	3.40	3.50
Situational awareness	4.00	3.40	3.40	3.40

collaboration being achieving the best score in the economy group, communication and data exploration being the strongest point for the IT group, and situational awareness leaving the greatest impact on the focus group. While a given bias towards a specific concept shifts the main purpose of the evaluated implementation into differing directions, the outcome provides a prospect regarding the individual hierarchy of interests of each user group in an inter-group setting.

8.5 Conclusions

This chapter presented an analysis of additional supporting concepts to be incorporated into design decisions regarding collaborative systems for raising situational awareness in the context of cybersecurity. Using the conducted workshops for CS-AWARE, as well as a state-of-the-art analysis regarding cybersecurity visualizations, a requirements analysis about functional and visual design specifications was created to support the extended concepts presented utilizing the CS-AWARE system architecture framework in its core.

The adapted components provide an overview of applied changes in terms of functionality and information flows. Major additions lie in the shifted focus of data exploration and intra-group communication, the latter of which can be extended into an inter-group setting encompassing additional domains through different projections of data meeting the interests of individual departments.

In order to facilitate the correlation of the effects of the supporting concepts and requirements on potential users, a prototype testbed was created to evaluate their interactions with the system. The evaluation process was divided into three parts consisting of usability tests, qualitative semi-structured interviews, and quantitative interviews to assess strengths and weaknesses of the implementation. The results provide an insight about the discrepancy between employed software tools in industry and state-of-the-art solutions.

In terms of visualization techniques used, nearly all participants preferred simple chart types with more complex types being optional on an individual level but detrimental for including other (potentially unrelated) users as the same level of comprehensibility cannot be guaranteed. While a higher degree of interactivity might inhibit intuitiveness, the evaluation results displayed a strong favour more interactive features, even if a short introduction in terms of usability was needed.

References

Aigner, W., Miksch, S., Schumann, H., & Tominski, C. (2011). *Visualization of time-oriented data.* Springer. https://doi.org/10.1007/978-0-85729-079-3

Bangor, A., Kortum, P. T., & Miller, T. J. (2008). The system usability scale (SUS): An empirical evaluation. *International Journal of Human-Computer Interaction, 24,* 574. https://doi.org/10.1080/10447310802205776

Brooke, J. (1995). *SUS: A quick and dirty usability scale* (p. 189). Usability Eval Ind..

Franklin, L., Pirrung, M., Blaha, L., Dowling, M., & Feng, M. (2017). Toward a visualization-supported workflow for cyber alert management using threat models and human-centered design. In *2017 IEEE Symposium on Visualization for Cyber Security (VizSec)* (pp. 1–8). https://doi.org/10.1109/VIZSEC.2017.8062200

Gonzalez-Granadillo, G., Alvarez, E., Motzek, A., Merialdo, M., Garcia-Alfaro, J., & Debar, H. (2016). Towards an automated and dynamic risk management response system. In B. B. Brumley & J. Röning (Eds.), *Secure IT systems* (pp. 37–53). Springer International Publishing. https://doi.org/10.1007/978-3-319-47560-8_3

Gonzalez-Granadillo, G., Dubus, S., Motzek, A., Garcia-Alfaro, J., Alvarez, E., Merialdo, M., Papillon, S., & Debar, H. (2018). Dynamic risk management response system to handle cyber threats. *Future Generation Computer Systems, 83*, 535–552. https://doi.org/10.1016/j.future.2017.05.043

Keim, D. A. (2002). Information visualization and visual data mining. *IEEE Transactions on Visualization and Computer Graphics, 8*(1), 1–8. https://doi.org/10.1109/2945.981847

Luidold, C. (2021). An architectural model for the support of dynamic risk management in an interdisciplinary setting [master, uniwien]. https://othes.univie.ac.at/68393/.

Luidold, C., & Schaberreiter, T. (2020). A gap analysis of visual and functional requirements in cybersecurity monitoring tools. In *SECURWARE 2020: The Fourteenth International Conference on Emerging Security Information, Systems and Technologies* (pp. 8–15).

Mattermost | Open Source Collaboration for Developers. (2015, June). Mattermost.Com. https://mattermost.com/

Mckenna, S., Staheli, D., & Meyer, M. (2015). Unlocking user-centered design methods for building cyber security visualizations. In *2015 IEEE Symposium on Visualization for Cyber Security (VizSec)* (pp. 1–8). https://doi.org/10.1109/VIZSEC.2015.7312771

Wagner, C., Dulaunoy, A., Wagener, G., & Iklody, A. (2016). MISP: The design and implementation of a collaborative threat intelligence sharing platform. In *Proceedings of the 2016 ACM on Workshop on Information Sharing and Collaborative Security* (pp. 49–56). https://doi.org/10.1145/2994539.2994542

Wagner, M., Fischer, F., Luh, R., Haberson, A., Rind, A., Keim, D., & Aigner, W. (2015, May 29). A survey of visualization systems for malware analysis.

Chapter 9
Other Applications for Cybersecurity Awareness

Jerry Andriessen, Thomas Schaberreiter, Alexandros Papanikolaou, Christopher Wills, and Juha Röning

9.1 The CS-AWARE Approach

The H2020-funded CS-AWARE project aims to equip local public administrations with a toolset allowing them to gain a better picture of vulnerabilities and threats or infiltrations of their ICT systems. This will be achieved via an underlying information flow model including components for information collection, analysis and visualisation which contribute to an integrated awareness picture that gives an overview of the current status in the monitored infrastructure and raises the awareness for both looming and already materialized threats (Schaberreiter et al., 2019).

While the systems that we built, with significant user input, can be deployed in similar contexts, i.e., other municipalities, in the current chapter, we envisage other domains or contexts of application.

J. Andriessen (✉)
Wise & Munro, The Hague, The Netherlands
e-mail: jerry@wisemunro.eu

T. Schaberreiter
CS-AWARE Corporation, Tallinn, Estonia
e-mail: thomas.schaberreiter@cs-aware.com

A. Papanikolaou
INNOSEC, Thessaloniki, Greece
e-mail: a.papanikolaou@innosec.gr

C. Wills
Caris Partnership, Fowey, UK
e-mail: chris.wills@cs-aware.com

J. Röning
University of Oulu, Oulu, Finland
e-mail: Juha.Roning@oulu.fi

© The Author(s), under exclusive license to Springer Nature Switzerland AG 2022   207
J. Andriessen et al. (eds.), *Cybersecurity Awareness*, Advances in Information
Security 88, https://doi.org/10.1007/978-3-031-04227-0_9

CS-AWARE is not only a toolset, it is also an approach to cybersecurity awareness. The following characteristics cover what we would like to call the CS-AWARE approach:

1. *Sociotechnical premise:* In the CS-AWARE approach we assume mutual constitution of people and (digital) technology. People and the technology interact in not deterministic ways, – meaning not dependent on surrounding events -, so we do well to study this interaction in the context of actual tasks for which people use the technology (Sawyer & Jarrahi, 2014; Tchounikine, 2016). The sociotechnical premise applies to how we conceive *the use* of the technology in context, but also to how we *design* the technology and its future use. This is realised through extensive user involvement during design workshops that apply the soft systems methodology (see Chap. 2 for more details). It implies that processes and tools are developed from the requirements analysis upwards.

2. *Systemic consideration:* Because the use of technology is socially situated, we have to take this context into account as much as possible. Understanding the entirety of the cybersecurity relevant aspects of the internal system is one of the cornerstones for ensuring useful as well as successful collaboration and cooperation between institutions. This means that our socio-technical analysis identifies the assets and dependencies within the system, captures the business processes that make use of the system, and includes the organisational roles, processes, and relations that would make for an effective cybersecurity monitoring and threat mitigation process. See Chap. 2 for more details about systems thinking, and Chap. 4 for the technological approach to critical infrastructure modelling (Schaberreiter et al., 2013) and visualisation of results (Eronen & Röning, 2006).

3. *EU-framework compatible*: cybersecurity is strongly interlinked with organisational and behavioural aspects of IT operations, and the need to adhere to the existing and upcoming legal and regulatory framework for cybersecurity. This is particularly true in the European Union, where substantial efforts have been made to introduce a comprehensive and coherent legal framework for cybersecurity. More details about the EU-framework are provided in Chap. 1 of this book.

4. *Collaborative understanding*: One of the key aspects of the European cybersecurity strategy is a cooperative and collaborative understanding of cybersecurity. CS-AWARE is built around this concept and relies on cybersecurity information being shared by relevant authorities in order to enhance awareness capabilities. At the same time, CS-AWARE enables system operators to share incidents with relevant authorities to help protect the larger community from similar incidents. A third aspect of collaboration refers to the way cybersecurity is managed and shared within the organisation itself (Andriessen & Pardijs, 2021).

5. *Multidisciplinary input*: Building a CS-AWARE solution requires input from many experts, with various backgrounds, academic as well as not academic. Our team includes technology experts, social scientists and experts with a background in humanities, especially law and language. In addition, it needs input from various professionals within the organisation, who are the experts in their domain.

The following sections present potential application areas for the CS-AWARE approach that are currently actively investigated.

9.2   Knowledge Management

Knowledge management (KM) is recognised as being a key issue in the management and maintenance of a huge range industrial, scientific, administrative and technical processes, including I.T. For example, KM is embodied in ITIL (White & Greiner, 2019), a structured approach to IT management which underpins ISO/IEC 2000. While the challenge of KM is pressing for all kinds of organisations, approaches have been developed to create tools which can search the sum of human knowledge, IBM 's Watson Discovery (Daws, 2021), an advanced question-answering system being an example.

At an organisational level, as processes and the technology that underpins them, become ever more complex, so does the complexity involved in organisations attempts to capture, store and access the knowledge required for operations. While capturing and retaining explicit knowledge is relatively straightforward, and can be embodied in technical manuals and user guides, capturing the knowledge created through the experience of operatives and employees, is much more problematic. A situation neatly captured by Nonaka and Takeuchi (1995), in their model of knowledge creation, depicted in Fig. 9.1.

- Socialisation in the above model, involves the transfer of tacit knowledge of an individual to the tacit knowledge of another individual takes place through

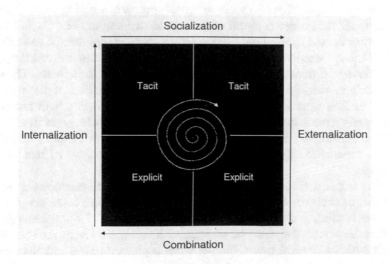

Fig. 9.1 The SECI model of knowledge creation

socialization; through practice observation and imitation. An example would be a master craftsman teaching an apprentice.

- Externalization is where tacit knowledge is made explicit and codified into some sort of record or documentation such as a manual. The problem with this is that tacit knowledge can be very difficult to codify since it may be innate and instinctive behaviour.
- Combination, where explicit knowledge in combination with other items of explicit knowledge is codified into a document to create new knowledge.
- Internalization involves the transformation of explicit knowledge into tacit knowledge. Explicit sources of knowledge are learned and internalised, at which point it becomes tacit knowledge.

As is clear from the above, the externalisation of tacit knowledge and its transformation into explicit knowledge is in many instances. How does a master carver know how to follow the grain of a stone in order to obtain the desired result? S/he does so through application of his/her innate understanding of the material with which he/she has worked over a period of many years. Such innate understanding and knowledge if very difficult, in some case impossible to codify transform into explicit knowledge. This sits at the heart of the KM problem. In one example from one of the author's experience a mass transit operator realised that over a period of the following 18 months several engineers responsible for maintaining the multiple types of metro station ticket gate used by the company were retiring. As they retired, each engineer's tacit knowledge about the maintenance and repair of the older types of ticket gate was lost to the company. Worked was commenced to attempt to codify the tacit knowledge of the engineers – a project that was both complex and long!

In Chap. 2, we described how the Systems Dependency Analysis (SDA) workshops were used in a holistic analysis of both an Larissa and Rome's cybersecurity situations, and of their information system architectures. These workshops were based upon the application of a socio-technical approach, working with the participants to facilitate the development of their understanding of their City's I.T. systems, networks, and applications. The approach taken complimented and mirrored the SECI model described above. The participants socialized in the workshops, and this facilitated the transfer of tacit knowledge between the participants. Their now shared tacit knowledge, was captured through the generation of Rich Pictures (RP's). The RPs were externalized, codified and combined into both textual and diagrammatic representations and records of their joint knowledge and understanding of the overarching problem domain. We presume that this explicit knowledge was then internalized by the participants, and retained by each of them as tacit knowledge.

Further to this, and at an operational level, the CS-AWARE platform captures the analytical and decision-making skills of those operatives who react to threats identified by the platform and externalises it by creating a report of it. It does so through recording the nature of a particular threat, and the actions taken by the operative to remediate and resolve the threat. Further work is planned to use both this, and the approach taken in the workshops, as the basis of a KM system that will be very widely applicable to a huge range of problem domains.

9.3   Better Cybersecurity Cooperation and Collaboration Between Different Organisations

Collaboration at the level of European legislation has always been crucially important, it is the core of European existence. Cybersecurity is no different. CS-AWARE was designed around the multi-level collaborative European cybersecurity framework as defined by the European cybersecurity strategy of 2013 and subsequent legislation, e.g. the Network and Information Security (NIS) directive. Chapter 1 describes these strategies in more detail.

Local supply chains are a significant factor for the smooth operation of industry and society: a cyberattack in one organisation may have disruptive cascading consequences on other organisations through supply chain dependencies. Unfortunately, there currently (2021) is no strategic approach to include organizations on the local/ regional level to actively participate in the collaborative European cybersecurity efforts.

To achieve such a goal focus needs to shift from purely intra-organisational cybersecurity considerations to the cooperation and collaboration on the local/ regional level, aiming for a more effective shared situational awareness in incident and risk management within the local/regional supply chain.

> There is a subtle distinction of meaning between the terms "collaborate" and "cooperate", although they both mean working ("labore") or "operating" together (the "co-" prefix). To cooperate with someone, means to perform certain actions that go towards achieving the other's goals, such as installing a security patch. But to collaborate, means to go much further: it is to actively work on elaborating a shared system of ideas and values. Collaboration can therefore be distinguished from cooperation in that in the case of collaboration we focus on the participants' joint activities characterised by a shared purpose and direction (Andriessen & Baker, 2020, p. 4).

In collaboration, all participants contribute more or less equally, according to their roles and expertise (these can be formally established or negotiated) and have equal status. To improve awareness of cybersecurity at local/regional levels we need to focus on collaboration in addition to cooperation.

The state-of-the-art approach to enable cooperation and collaboration, predominantly on the national/European NIS sectoral level, are information sharing and analysis centres (ISACs). ISACs are meant to promote and regulate (especially) information sharing through co-regulation (the regulatory body gives power and entrusts market stakeholders to achieve a policy objective). A review study by ENISA (European Network and Information Security Agency, 2017) revealed many ISACs being established and co-regulatory initiatives are increasingly expanding and grow in popularity, but they still have to address a number of challenges: lack of (or weak) enforcement and trust, the voluntary nature of membership and, quite often, the voluntary (or not really monitored) nature of information sharing. Moreover, ISACs currently do not (yet) operate at local/regional levels.

A tentative proposal for a cybersecurity cooperation and collaboration framework for the local/regional level could extend and improve the ISAC model to be able to operate in a more structured and dynamic way, by better integrating cooperative and collaborative procedures with the dynamic cybersecurity risk and incident management tools and procedures that the next generation of cybersecurity tools like CS-AWARE offer. We could for example achieve this within the CS-AWARE approach by:

- Defining a socio-technical methodology for capturing and visualizing cybersecurity interdependencies at multiple layers (infrastructures, data, vulnerabilities, services for citizens, regulations, partners and roles, business processes, human behaviour and its possible impact). This will be based on the soft systems methodology (SSM) based approach developed during the CS-AWARE project, which will be extended from capturing the organizational context to be able to capture the local/regional context as well.
- Building a framework for facilitating cooperation/collaboration based on this shared situational awareness to incentivise organizations to share information and cooperate/collaborate on cybersecurity problems that go beyond the organizational context. This will be a shared effort between all stakeholders (end users, competent authorities/CSIRTs and academic/industry partners).
- Providing a collaboration infrastructure to support the efforts, and integrate the collaboration with organizational cybersecurity risk and incident management. Currently it is common that communication through email or telephone happens in the case of incidents, but this is often coordination between individuals. Collaboration does not simply happen by requiring it. For communication to become collaborative, it needs to be based on evolved trust and equality, on joint agreement about the things that should be addressed, in the context of mutual respect and consideration. This can be facilitated through an intuitive and easy to use support infrastructure. For example, the SPOD platform (Hogan et al., 2017; Scarano et al., 2017) developed in the context of the ROUTE-TO-PA project (Ojo et al., 2018) has a proven track record for supporting collaboration and transparency in the public sector. It can be used as a basis and adapted to the specific cooperative and collaborative needs of cybersecurity.

9.4 Smart City Applications

One of the major emerging technologies in the context of LPAs is without a doubt the continuing digitization of administrative services commonly referred to as "smart city". For example, our partner city Rome has a wide ranging and long-term strategy of implementing smart city capabilities in the context of the Roma Data Platform (RDP).

Considering the growing demand for innovative IT tools, Roma Capitale decided to customize and implement a Smart City Strategy aims at increasing efficiency and effectiveness of interactions with citizens as well as monitoring the level of services provided. The smart city strategy will be implemented incrementally, starting with various applications in Education, Taxation and Reports/Complaints, but aims at integrating virtually all administrative areas over time.

The RDP at the core of the strategy is the central hub for collecting data, processing data and distributing it to end user applications. The platform has following core responsibilities:

- RDP is an IT Data Platform aims to data integration between Administration and external data sources and to improve **data-driven governance policies**.
- RDP offers built-in, ready to use services but also functions that can be integrated into other applications.
- RDP supports ecosystems' growth.
- RDP is an enabler of Public Private Partnership models for data management.

In general, the increasing ability and affordability to collect, process and store increasing amounts of data collected from a multitude of sensing and collection devices in a multitude of use case scenarios allows to process and fuse the data to be applicable to the needs of a wider range of end user groups than is served today by digital administration.

While the scope of the wide-ranging smart city efforts of the city of Rome certainly have a dimension that only metropolitan areas can cope with, the continuing digitization of services and growing availability of data will also compel medium and small sized municipalities to implement smart city applications. For example, our partner city of Larissa – as an example of a medium sized municipality - has recently implemented city wide smart infrastructure projects, and has plans for further applications.

From a security perspective, smart city applications will necessitate a stronger interconnection of enabling technologies like for example IoT, 5G connectivity, AI, smart traffic applications to collect and process relevant data. Furthermore, the volume of data that will be collected will increase significantly, and the data will need to be more personal for many smart city applications to work.

Aside from that, the general service model of data collection, data processing, data storage and data visualization/utilization will follow the same principles as it does today.

Given those circumstances, the CS-AWARE approach is very well suited to deal with the cybersecurity awareness requirements of emerging smart city applications. The identification of assets, dependencies, information flows and business processes and monitoring points – and how those aspects relate to the cybersecurity state of the application – is the first step in any new application that is monitored by CS-AWARE. Due to flexibility that the elicitation of the individual application/service requirements brings, and the way we translate those aspects into monitoring patterns, CS-AWARE is entirely ready for the smart city future!

9.5 e-Democracy and e-Governance in the EU Due to COVID-19

The COVID-19 pandemic created an increased need for remote working, which in turn demanded for an extensive use of Information and Communications Technology (ICT). For the vast majority of employees this meant that they should be given access to the systems they require for carrying out their everyday tasks. Similarly, Officials had to perform any decision-making processes (e.g. Council Boards) in a fully electronic way. From the citizens' point of view, several services they were used to access in a physical manner would now have to be provided electronically. In order to achieve this state, several technological solutions were employed extensively, one of which being the Virtual Private Network (VPN). This allowed employees to securely connect to their organisation's internal network and carry out their work.

The notion of e-Democracy has been around for quite some time and involves the use of ICT to support the democratic decision-making processes (Macintosh 2004). Another related notion, that of the e-Government, concerns the application of IT for delivering government services, exchange of information, communication transactions, integration of various stand-alone systems between government to citizen (G2C), government-to-business (G2B), government-to-government (G2G), government-to-employees (G2E) as well as back-office processes and interactions within the entire government framework.

For most EU countries, the COVID-19 pandemic triggered a transition to a fuller implementation of both e-Democracy and e-Government. For countries that had a relatively small and/or partial implementation of e-Democracy and/or e-Government, the transition turned out to be quite a violent and fast-paced one. What is more, the cost of this transition is not to be neglected, since a significant investment was made on hardware and/or online services and this equipment is therefore expected to continue being used quite extensively. In addition, the economic impact of the COVID-19 pandemic is expected to force them on budget cut-downs in the near future and hence their only choice may be to continue their operation remotely, via electronic services.

Within this domain, the CS-AWARE approach can be applied productively. As far as Local Public Administrations (LPAs) are concerned, there are cases where a council board concerning a large administrative region has to be held and representatives from smaller LPAs (yet belonging to the same administrative region) have to participate. This is therefore expected to lead to having many more remote users (e.g. via VPN or teleconference software), which are harder to manage and monitor for cybersecurity issues. For instance, the location where each user is connected from may be an initial indicator regarding a potential breach of confidentiality that crosses the country's borders. Another example could be a malicious participant who is trying to gain access to information and systems they are not supposed to. Given the increased complexity and the large number of users, cybersecurity awareness would be extremely valuable from the IT administrators point of view, as it would give them in real-time the wider picture of what is happening in the information system, they are responsible for.

9.6 Managing Cross-Sector and Cross-Border Dependencies in the European Critical Infrastructure Context

Starting with the European cybersecurity strategy of 2013 (*Towards a More Secure, Global and Open Cyberspace*), Europe is committed to a cooperative/collaborative approach to cybersecurity in order to be able to better protect essential services from incidents that threaten society and economy, while facing a constantly changing threat landscape. Subsequent legislation like the 2016 Network and Information Security (NIS) directive (Smith, 2015) (energy, health, transport, banking, digital infrastructure, financial market infrastructures, drinking water supply and distribution, digital service providers) and the 2018 European Electronic Communications Code (European Parliament, 2018) for the telecommunications sector have specified the framework and defined and mandated the rights and obligation of the relevant actors and stakeholders. This framework includes the obligation for individual operators of critical services to collaborate on risk assessments on a regular basis with competent authorities (CAs) established by member states on a national level (NIS/EECC CAs).

While this approach allows CAs to gain insight into individual operator risks, and ideally draw conclusions about risks on a sectoral level, systemic aspects including large-scale cross-sectoral and cross-border risk, weaknesses and vulnerabilities cannot be adequately assessed following this approach. This shortcoming has been a major finding published in the updated European cybersecurity strategy of 2020 (European Commission, 2020b), and is addressed in the proposed update to the NIS directive, the NIS2 (European Parliament, 2020b), calling for a stronger cross-sector and cross-border collaboration to address this issue. Similarly, the proposal directive on the resilience of critical entities (European Parliament, 2020a), which aligns the European framework for dealing with physical risks with the framework for cyber risks defined by the NIS directive, is designed with cross-sector and cross-border risks in mind as well.

Effectively, NIS/EECC CAs (and in future CAs implementing the resilience directive) are the only European authorities that have on the one hand the legal mandate to deal with, and on the other hand the required data/intelligence available to assess risk of European critical infrastructure, enabling them in principle to manage large-scale systemic risk in cooperation with the individual operators. However, in order to include systemic risk considerations, additional requirements to the currently implemented approach need to be fulfilled, since CAs currently do not have the capabilities to complement individual operator risk assessments with systemic aspects.

In order to achieve this, a cooperative/collaborative approach is required to assess the systemic aspects that go beyond individual operators or sectors. In this context, the CS-AWARE cybersecurity awareness approach is well suited to facilitate the identification, awareness and visualisation of complex dependencies between operators and sectors, even across national borders. The soft-systems based analysis methodology (SSM) developed during the CS-AWARE project for analysing and visualising cybersecurity aspects in complex organisational settings can be easily adapted to the inter-organisational, cross-sector and cross-border context that is so crucial to understand the risks, weaknesses and vulnerabilities associated to the systemic aspects of European critical infrastructure interdependency. This will provide CAs with the necessary capabilities and data for managing the systemic aspects related to cybersecurity of European critical infrastructure, allowing for more accurate risk assessments on this level, and provide a basis for collaboration on preparedness, resilience and response in order to achieve a high level of common cybersecurity across the Union, as designated by the NIS directive.

9.7   Autonomous Collaborative Robots: UAV, UGV, Water Vessels Heavy Duty Machines

The robotization of industry is one of the key drivers behind the Industry 4.0 revolution (Interreg Europe, 2019). Currently, collaborative robots are becoming a reality across the manufacturing and other industries,[1] autonomous robots are already a key asset in the logistics sector, and unmanned aerial vehicles (UAVs) are being used for inspection and monitoring in diverse domains. Ubiquitous robots with augmented connectivity are merging into the Industrial Internet of Things (IIoT),[2] enabling deeper degrees of intelligence through computational offloading. Multiple challenges constrain further innovation in these areas: higher-throughput and lower-latency wireless networks with more robust security solutions from the connectivity point of view (Latva-aho & Leppänen, 2019), and long-term autonomy together with more efficient, distributed, and autonomous multi-robot collaboration from the robotics side (Vaidya et al., 2018). In recent years, developers and robotics experts have created the first pieces of autonomous construction equipment. They will soon become part of highly complex systems that can automate some or all the work on construction sites. The autonomous construction equipment market is valued at approximately US $9.53 billion (2019).[3]

---

[1] https://www.interactanalysis.com/the-collaborative-robot-market-2019/.

[2] https://www.ericsson.com/en/reports-and-papers/white-papers/5g-wireless-access-an-overview.

[3] https://www.globenewswire.com/news-release/2020/08/05/2073458/0/en/The-Global-Autonomous-Construction-Equipment-Market-Report-2020-Delves-Into-The-Autonomous-Construction-Equipment-Industry-s-Recovery-After-The-COVID-19-Crisis.html.

The trend of investing in mobile robotics in the manufacturing plants and construction works increasing day by day from the companies to make their production construction work more efficient, safer and agile. The Industry 4.0 approaches can function optimally when production and logistic processes are fluently connected and linked with each other, without resulting idle times for restoring components. At the same time data, connectivity and reliability need for all operations is rising. We have seen symptoms of malware affecting production machinery. Wannacry stopped two large rock drilling machines in Kevitsa[4] Especially on construction side continuous flow from cloud to a machine and back is needed. This is usually arrange using BIM approach.

Building Information Modeling (BIM) is the holistic process of creating and managing information for a built asset. Based on an intelligent model and enabled by a cloud platform, BIM integrates structured, multi-disciplinary data to produce a digital representation of an asset across its lifecycle, from planning and design to construction and operations.

Autonomous aerial logistics solutions are at the intersection of today's digitalization of industry, transformation of urban design, development of new power solutions for transport, and at the helm of the technological transitions of society. These autonomous systems are set to play an essential role in climate change mitigation and enable long-term sustainability and efficient services globally in the increasingly urbanized world. Multiple challenges are constraining further innovation and wider adoption of green and sustainable aerial urban logistics: the lack of the integration to the urban landscape (Otto et al., 2018); the higher degree of coordination between air and ground units (Drone Alliance Europe, 2020; EASA, 2020); the trustworthiness and interoperability of digital solutions (Azzoni, 2020)[5]; the autonomy of the drones and the supporting infrastructure (EASA, 2021; European Commission, 2020a); and the energy density limitations of current energy solutions (Blondelle & Research, 2021; Wang et al., 2020).

The world is progressing in the direction of automation in its many different configurations. For this seamless combination of data, AI and mechatronics equipment are required. From a single machine to fleet their operation need to be monitored in trusted way. CS-AWARE can bring understanding for the architecture and its interfaces. There will be many players in this, public, private, professionals and even amateurs which makes it harder to manage and monitor for cybersecurity issues. A drone flying from a city area to another one needs to seamlessly switch to another UTM, have probably access to cloud based information and be monitored and have to report and log its operations. Putting multiple of them and allowing further collaboration with other kinds of mobile robots and machines makes it even more challenging.

CS-AWARE will work to provide cybersecurity when developing a model to manage sharing and use of data, including federated data access and management.

---

[4] https://www.iltalehti.fi/digi/a/201705172200143927.

[5] https://beta.oulu.fi/en/projects/robomesh.

This work is done considering and aligning ongoing efforts in European wide e orts such as GAIA-X, working towards a federated and secure data infrastructure, and those of International Data Spaces Association (IDSA).

9.8 Conclusion

In this final chapter, we proposed a number of areas in which the CS-AWARE approach could provide an important contribution to cybersecurity. Because the number of possible application areas appears to be very large, and the imminent need for a systematic approach to cybersecurity awareness, we decided to continue our collaboration. CS-AWARE will continue as a company with a promising business plan. In addition, we are proposing several EU-grants for further extension of our reach and further elaboration of our market. And, finally, our multidisciplinary collaboration team plans further interdisciplinary publications for a variety of target groups.

References

Andriessen, J., & Baker, M. (2020). On collaboration: Personal, educational and societal arenas.. Brill Sense.

Andriessen, J., & Pardijs, M. (2021). Awareness of cybersecurity: Implications for learning for future citizens. In Z. Kubincová, L. Lancia, E. Popescu, M. Nakayama, V. Scarano, & A. B. Gil (Eds.), *Methodologies and intelligent Systems for Technology Enhanced Learning, 10th international conference. Workshops* (Vol. 1236, pp. 241–248). Springer International Publishing. https://doi.org/10.1007/978-3-030-52287-2_24

Azzoni, P. (2020). *From internet of things to system of systems [an Artemis-IA whitepaper].* Artemis Industry Association.

Blondelle, J., & Research, D. (2021). *European framework for power-to-X* (p. 19). DG Research and Innovation.

Daws, R. (2021, November 10). IBM enhances Watson Discovery's natural language processing. AI News. https://artificialintelligence-news.com/2021/11/10/ibm-enhances-watson-discovery-natural-language-processing-capabilities/

Drone Alliance Europe. (2020). U-space Whitepaper Version 2.0. https://resourcecenter.dronealliance.eu/wp-content/uploads/2019/05/DAE-UTM-U-Space-whitepaper-2.0-final-1.pdf

EASA. (2020). *Draft acceptable means of compliance (AMC) and guidance material (GM) to opinion no 01/2020 on a high-level regulatory framework for the U-space.* European Union Aviation Safety Agency. https://www.easa.europa.eu/official-publication/acceptable-means-of-compliance-and-guidance-materials

EASA. (2021). *Study on the societal acceptance of urban air mobility in EuropeMay [this study has been carried out for EASA by McKinsey & Company].* European Union Aviation Safety Agency. https://www.easa.europa.eu/sites/default/files/dfu/uam-full-report.pdf

Eronen, J., & Röning, J. (2006). Graphingwiki—A semantic wiki extension for visualising and inferring protocol dependency. In M. Völkel (Ed.), *Proceedings of the First Workshop on Semantic Wikis – From Wiki To Semantics* (pp. 1–16).

European Commission. (2020a). *Sustainable and Smart Mobility Strategy – putting European transport on track for the future* (COM/2020/789 final). https://eur-lex.europa.eu/legal-content/EN/TXT/?uri=CELEX%3A52020DC0789

European Commission. (2020b). *The EU's Cybersecurity Strategy for the Digital Decade | Shaping Europe's digital future.* https://digital-strategy.ec.europa.eu/en/library/eus-cybersecurity-strategy-digital-decade-0

European Network and Information Security Agency. (2017). *Information sharing and analysis centres (ISACs): Cooperative models.* Publications Office. https://data.europa.eu/doi/10.2824/549292

European Parliament. (2018). *Directive (EU) 2018/1972 of the European Parliament and of the council of 11 December 2018 establishing the European electronic communications code (recast)text with EEA relevance.* p. 179.

European Parliament. (2020a). DIRECTIVE OF THE EUROPEAN PARLIAMENT AND OF THE COUNCIL on the resilience of critical entities.

European Parliament. (2020b). *Proposal for a DIRECTIVE OF THE EUROPEAN PARLIAMENT AND OF THE COUNCIL on measures for a high common level of cybersecurity across the union, repealing directive (EU) 2016/1148* (p. 12).

Hogan, M., Ojo, A., Harney, O., Ruijer, E., Meijer, A., Andriessen, J., Pardijs, M., Boscolo, P., Palmisano, E., Satta, M., Groff, J., Baker, M., Détienne, F., Porwol, L., Scarano, V., & Malandrino, D. (2017). Governance, transparency and the collaborative design of open data collaboration platforms: Understanding barriers, options, and needs. In *Government 3.0 – Next Generation Government Technology Infrastructure and Services* (pp. 299–332). Springer. https://doi.org/10.1007/978-3-319-63743-3_12

Interreg Europe. (2019). *Industry 4.0: A Policy Brief* (A Policy Brief from the Policy Learning Platform on Research and Innovation). Policy Learning Platform on Research and innovation. https://www.interregeurope.eu/fileadmin/user_upload/plp_uploads/policy_briefs/INDUSTRY_4.0_Policy_Brief.pdf

Latva-aho, M., & Leppänen, K. (2019). Key drivers and research challenges for 6G ubiquitous wireless intelligence. In *6G flagship*. University of Oulu.

Nonaka, I., & Takeuchi, H. (1995). *The knowledge-creating company: How Japanese companies create the dynamics of innovation.* Oxford University Press.

Ojo, A., Stasiewicz, A., Porwol, L., Petta, A., Pirozzi, D., Serra, L., Scarano, V., & Vicidomini, L. (2018). *A comprehensive architecture to support open data access, co-creation, and dissemination.* pp. 1–2. https://doi.org/10.1145/3209281.3209411.

Otto, A., Agatz, N., Campbell, J., Golden, B., & Pesch, E. (2018). Optimization approaches for civil applications of unmanned aerial vehicles (UAVs) or aerial drones: A survey. *Networks, 72*(4), 411–458. https://doi.org/10.1002/net.21818

Sawyer, S., & Jarrahi, M. H. (2014). Sociotechnical approaches to the study of information systems. In *Computing Handbook* (3rd ed., pp. 5-1–5–27). CRC Press. https://doi.org/10.1201/b16768

Scarano, V., Malandrino, D., Baker, M., Détienne, F., Andriessen, J., Pardijs, M., Ojo, A., Hogan, M., Meijer, A., & Ruijer, E. (2017). Fostering citizens' participation and transparency with social tools and personalization. In *Government 3.0 – Next generation government technology infrastructure and services* (pp. 197–218). Springer. https://doi.org/10.1007/978-3-319-63743-3_8

Schaberreiter, T., Kittilä, K., Halunen, K., Röning, J., & Khadraoui, D. (2013). Risk assessment in critical infrastructure security modelling based on dependency analysis. In S. Bologna, B. Hämmerli, D. Gritzalis, & S. Wolthusen (Eds.), *Critical information infrastructure security* (pp. 213–217). Springer. https://doi.org/10.1007/978-3-642-41476-3_20

Schaberreiter, T., Roning, J., Quirchmayr, G., Kupfersberger, V., Wills, C., Bregonzio, M., Koumpis, A., Sales, J. E., Vasiliu, L., Gammelgaard, K., Papanikolaou, A., Rantos, K., & Spyros, A. (2019). A cybersecurity situational awareness and information-sharing solution for local public administrations based on advanced big data analysis: The CS-AWARE Project. In J. B. Bernabe & A. Skarmeta (Eds.), *Challenges in Cybersecurity and Privacy – the European Research Landscape* (pp. 149–180). River Publishers.

Smith, R. (2015). Directive 2010/41/EU of the European Parliament and of the Council of 7 July 2010. In R. Smith (Ed.), *Core EU legislation* (pp. 352–355). Macmillan Education. https://doi. org/10.1007/978-1-137-54482-7_33

Tchounikine, P. (2016). Contribution to a theory of CSCL scripts: Taking into account the appropriation of scripts by learners. *International Journal of Computer-Supported Collaborative Learning, 11*(3), 349–369. https://doi.org/10.1007/s11412-016-9240-8

Vaidya, S., Ambad, P., & Bhosle, S. (2018). Industry 4.0 – A glimpse. *Procedia Manufacturing, 20*, 233–238. https://doi.org/10.1016/j.promfg.2018.02.034

Wang, B., Zhao, D., Li, W., Wang, Z., Huang, Y., You, Y., & Becker, S. (2020). Current technologies and challenges of applying fuel cell hybrid propulsion systems in unmanned aerial vehicles. *Progress in Aerospace Sciences, 116*, 100620. https://doi.org/10.1016/j.paerosci.2020.100620

White, S. K., & Greiner, L. (2019, January 18). What is ITIL? Your guide to the IT infrastructure library. CIO. https://www.cio.com/article/2439501/infrastructure-it-infrastructure-library-itil-definition-and-solutions.html

Printed in the United States
by Baker & Taylor Publisher Services